Uncommons in the Commons

About the Author

Rucha Ghate was awarded Ph.D in 1989 by Nagpur University on her thesis entitled "Forest Policy and its Economic Impact on Tribals of Maharashtra". It was subsequently published in a book form. Her interest in forests and indigenous people of Vidarbha led her to pursue the subject further after completing post doctoral fellowship in Environmental Economics from Indiana University, Bloomington, USA. She has undertaken and completed many research projects pertaining to the subject.

At present she is the Secretary of SHODH: The Institute for Research and Development, based in Nagpur and the Director of Collaborative Research Centre, IFRI. She has been a faculty member in the Post Graduate Teaching Department of the Nagpur University for the past 17 years, and has taught at Asian Institute of Technology (AIT), Thailand, as a visiting faculty, where she was awarded 'Teaching Award 2003'. She is a consultant to the World Bank and other international agencies for various projects. She has a book, many publications and numerous articles in her name.

UNCOMMONS IN THE COMMONS

Community Initiated Forest Resource Management

RUCHA GHATE

With Botanical inputs from **Alka Chaturvedi**

CONCEPT PUBLISHING COMPANY, NEW DELHI-110059

ISBN 81-8069-099-7

First Published 2004

Published and Printed by
Ashok Kumar Mittal
Concept Publishing Company,
A/15-16, Commercial Block, Mohan Garden,
New Delhi-110059 (INDIA)

Phones : 25351460, 25351794
Fax : 091-11-25357103
Email : publishing@conceptpub.com

Dedicated to the fond memory of my father who has been my first and greatest critic and hence a constant source of inspiration

FOREWORD

The importance of common pool resources, for the conservation of natural resources and survival of the communities dependent on them, is becoming more and more important with every passing day. For large number of socio-economically disadvantaged people, who either own no private resource or have very limited access to the same, the access to common pool/property resources and open access or public resources determines the difference between life and death. For majority of landless communities, the collection of biomass from such lands or water bodies provides the only source of livelihood. How these resources are managed depends on the institutional arrangements that have evolved over a period of time. Grafting new institutions on the existing ones, howsoever weak, requires understanding of and appreciation for the local cultural norms, community interactions and historically evolved patterns of resolving local conflicts.

I am extremely happy that the author has pursued case studies in this book using a robust framework and blending the communities' perspective with that of the outsiders. The study of five communities provides lessons not only about emergence of local leadership but also about the resolution of local conflicts. It is obvious that these conflicts need not always be resolved keeping the interest of the poor in mind. Local institutions have to operate within an informal cultural context but these are also influenced by the formal institutions. The study points out the enormous influence the Forest Department officials have had on the ability of communities to shape their institutional arrangements. The relationship between institutional arrangements and local ecological outcomes is another distinction of the study.

Given the paucity of well-documented empirical studies of local institutions providing the interface between informal culture and formal structures of local governance, this study makes a useful contribution. The botanical attributes of four villages demonstrate how significant is the influence of institutions on the ecological and botanical attributes. There is no doubt that role of local self-design institutions is only going to increase in the times to come. The real question is whether local institutions can withstand the destructive pressures created by the forces, sometimes beyond the control of the local communities. The role of networking among the communities and their leadership is another issue that remains to be explored. The evolution of polycentric society requires expansion of spaces for articulation and assertion of local leadership amenable to collective governance. This study provides useful insights about how this tension is likely to be resolved in future.

PROF. ANIL K. GUPTA
Professor
Indian Institute of Management
Ahmedabad 380015
India
anilg@sristi.org

PREFACE

For all of us who are interested in 'common pool resources' (CPRs) and 'communities' as academicians, social workers, or bureaucrats, it is a common experience to come across some communities that show positive synergy in solving their commons problem. At the same time we are intrigued by the fact that such communities seldom become role models for other neighbouring communities to follow. Why do the communities decide to come together in the first place? How is the act initiated? What keeps them going? And even if they positively, economically benefit from their endeavour, why don't the other communities follow their example?

It was this curiosity that led me to undertake a study of the communities who had initiated protection, conservation and in some cases even management of the forests surrounding their villages. During mid 80s, when I was engaged in studying the impact of subsequent forest policies on indigenous people of the State of Maharashtra, I came across the works of academia working around the problems of centrally owned and managed method of forest conservation. The general criticism since then has been moving around the limitations of the "keep people away from forests" principle adopted by the state. This principle, in practice, had turned the resource into a more or less 'open access' resource. The state had entrusted in itself the sole responsibility of guarding forests driven by the Hardinian rationality of 'in constraintless situation people at large would degrade the resource'.

However, most of the research in the field of CPRs, providing insight in commons dilemma, gives one hope that it is possible to build a theory of commons that will help to predict 'in which circumstances individuals or communities are more likely to craft new institutions capable of governing their commons'. The purpose of this book is to re-emphasise the fact that communities can and do work towards managing their resources, with or without the support of the state. At the same time, it also wishes to bring home the message that if the state can provide positive input to such efforts, the returns from community-managed efforts would certainly be high. For doing this, identifying local leadership, providing positive incentives, developing skills of people, and providing technical know-how, need to be supported either by the state or by Non Governmental Organizations (NGOs). It is a well-established fact today that no institutional arrangement can completely protect forests without taking the communities, residing around the resource, into confidence. Wherever communities are not made stakeholders, forests are bound to dwindle in the wake of growing population pressure and lack of available alternatives.

Unlike the reports and case studies from Rajasthan, Himachal Pradesh and of late from Andhra Pradesh and Madhya Pradesh, very little information is available regarding communities indulging in forest management in Maharashtra State in general and Vidarbha region in particular. This book attempts to throw light on two communities that brought revolution in their lives very silently. It all started more than a decade ago. However, it is only now that their efforts are being given some cognizance. Saigata and Mendha have set examples of how rational actors are capable of creating durable institutions that allow them in avoiding 'tragedies of commons' by engaging in collective action. Lakhapur community presents a different dimension of collective action that is influenced by the philosophy of 'total transfer of decision-making rights' to communities. Though it sounds rational if taken on its face value, its impact on the resource as well as on the people has been mixed. Village Lekha is an example of a heterogeneous community, with relative abundance of forests, struggling to come together in the changed circumstances. However due to indifferent Forest Department officials, and lack of strong leadership within the community, it has yet not achieved much.

Maral Mendha is an eye-opener. This community has strategically encroached on forest land, converting it into cultivated land, resulting in enhanced immediate incomes. It is an example of perfect collective action, but with short-run, self-centered interest in mind. These five communities are at three levels of collective action, each giving out lessons to learn.

It is generally believed that lack of clear land tenure discourages people from coming together. This book is also my appreciation of the communities who have invested their time and effort in protecting a resource as a social good, without tenure rights. The study being presented here is not meant to be a scholarly work in pursuit of academic excellence. It is meant for people working at the grass roots (or on grass top, as some like to call themselves), for communities that are still struggling to put their collective action together, and for policy makers who are searching for alternative ways of conserving the dwindling green resource.

My interest in CPRs was kindled at a Summer School organized by IIM, Ahmedabad in 1997. Dr. Anil Gupta, a total (almost passionate) believer in the capabilities of communities in solving their own common's problem inspired me to open my eyes towards such communities in Vidarbha region. I am privileged to have his Foreword to this book. The book is an outcome of a research project funded by the Workshop in Political Theory and Policy Analysis, Indiana University, USA. I am indebted to Dr. Elinor Ostrom not only for extending the funding, but also for developing my analytical skills and for providing very valuable inputs. I have very extensively quoted from her works. Dr. Amy Poteete gave me many constructive suggestions after going through the project report very meticulously. I was greatly benefited from my interactions with Dr. Clark Gibson, Dr. Catherine Tucker, and Dr. Arun Agrawal. Dr. Prakash Bobade, Dr. Vinayak Deshpande, Dr. Mohan Kashikar, Dr. Kishor Mahabal, Mr. Shree Bhagwan, Mr. Phani Kumar Garlapati, and Mr. Kailash Thawre were the team members of the research project. I gratefully acknowledge their contribution in the work. The research project was undertaken by the Collaborative Research Center (CRC) of International

Forestry Resources and Institutions (IFRI) Research Program, Indiana University, USA, placed in SHODH: The Institute for Research and Development, Nagpur. Mr. Mukund Kulkarni, President of SHODH, has been with me throughout. I am indebted to him for his useful suggestions and insight in rural psyche, which helped me interpret my observations from time to time. Dr. S .V. Khandewale, Dr. Vijay Vaidya, Dr. Anjali Kulkarni and Dr. Ramdas Mahore have always extended their support in my academic pursuits. Dr. S. G. Kane and Dr. Gopal Kadekodi provided their valuable guidance in the analysis of data. I acknowledge their support. Dr. Vishal Lichade helped us with data entry of the enormous information that was collected using the IFRI protocol. It is well known that a lot of hard work goes into sorting and feeding of data, literature search, maintenance of record, preparation of draft, editing etc. Ms. Deepshikha Mehra and Ms. Alpana Bose handled all these jobs in a very efficient manner. It would have been very difficult to put this work together without their untiring work.

All those working in the field of 'forests' know well that it is impossible even to access some villages without the active help of the Forest Department. In this research endeavour, I received help from many forest officials. Mr. Gurme, Deputy Conservator of Forests, Gadchroli division, and Mr. Khadse, Deputy Conservator of Forests, Chandrapur division, were very forthcoming in extending every possible help. Information provided by all the five communities is the base of the study being presented here. All the communities were very cooperative and patient with our interference in their daily routine. We were their guests for long periods, for more than a week at times, but they never made us feel unwelcome. Collection of forest data required laying of temporary plots and finding these random plots on maps, in the real forest. I wonder if I could have done this work without their help especially because topo-sheets were not available. Often three to four villagers volunteered to accompany us and actively got involved in forest mensuration. No words would suffice to acknowledge their contribution in making this study possible. I am especially thankful to Mr. Suryabhan Khobragade and

Mr. Devaji Tofa of Saigata and Mendha villages respectively for spending long hours with us, narrating the past and also for allowing me to use their names in this book. Involvement and patience of my family members is beyond the words of acknowledgement.

I am grateful to Nagpur University for extending funds for publication provided by the University Grants Commission, New Delhi. I am also indebted to the two anonymous referees for their recommendations.

RUCHA GHATE

Contents

List of Maps

List of Indian Words used in the Book

1.	*Andhari Van*	The dark Forest.
2.	*Awla*	*Emblica officinalis – (A Floral Specie)*
3.	*Bija*	*Pterocareus marsupium – (A Floral Specie)*
4.	*Char*	*Buchanania lanzan – (A Floral Specie)*
5.	*Dandaar*	Concise dance-dramas performed traditionally in Maharashtra villages.
6.	*Dhawda*	*Anogeissus latifolia – (A Floral Specie)*
7.	*Elge Aher*	The spot in the forest where sloth bears come for water
8.	*Gawatache Beed*	Forest patch set aside for grass
9.	*Gotul*	A traditional *Gondi* community centre.
10.	*Gram Panchayat/ Panchayat*	The lowest unit of local government
11.	*Gat-gram Panchayat*	Gram Panchayat for a group of small villages
12.	*Gramsabha*	A general body of the villagers
13.	*Gram-Swarajya*	Village autonomy
14.	*Gram-Swarajya Jungle Sanstha*	Forest association of a self-ruled village
15.	*Jaganeke*	A particular spot in the forest where the land emits a vibrating sound
16.	*Jangal Bachao – Manav Bachao Andolan*	Save forests-save human beings movement
17.	*Kosran*	The part of the forest a *Kos* (Local term for a particular distance) away from the village.
18.	*Kotwal*	The lowest official in the bureaucratic set-up for revenue collection.
19.	*Kotwal Raan*	Forest patch named after the *Kotwal*
20.	*Krishak Charcha Mandal*	Farmers Discussion Forums

21.	*Krishak Prashikshan*	Farmers' training
22.	Mahur Jar Palla	A spot in the forest where nothing but grass grows
23.	*Malgujar*	A person appointed by Government to collect revenue in British period.
24.	*Malgujari*	The system of collecting revenue
25.	*Mothya Devache Raan*	Forest of the "Superior" God (A sacred grove)
26.	*Mukaji Devache Raan*	Forest of the "Dumb" God
27.	*Moha*	*Madhuca longifolia – (A Floral Specie)*
28.	*Maharaja*	Emperor
29.	*Marar*	A caste of vegetable growers
30.	*Marathi*	Local language spoken by the natives of Maharashtra state
31.	*Nistar*	The rights of local residents over forest produce
32.	*Oop-Sarpanch*	Deputy to the *Sarpanch*
33.	*Pandal*	Tent put up for marriages and other community occasions.
34.	*Saheban Metta*	The *metta* (hill), which was once visited by a *Saheb* (An Englishman)
35.	*Sant*	Saint
36.	*Sarpanch*	Head of the village *Panchayat*
37.	*Tahangad Metta*	A high and huge hill
38.	*Tiwas van*	Forest with Tiwas (*Ougeinia oojeinensis)* as a predominant floral specie
39.	*Tulurmutha*	A traditional post-harvest ceremony practiced by the Gond tribe
40.	*Van Sanrakshan Samiti*	Forest Protection Committee
41.	*Wanar Chua*	A water body in the forest where monkeys come for water
42.	*Zamindar*	Local land holders who assist in collection of land cess
43.	*Zilla Parishad*	A government agency at district level

1

SELF-GOVERNANCE : WHY IS IT RARE ?

"Scholars of common property have shown that markets or private property arrangements and state ownership or management do not exhaust the range of plausible institutional mechanisms to govern natural resource use".

—Arun Agrawal

The unsustainable rate of forest exploitation in tropical regions is a matter of great concern for governments, environmentalists, and local communities alike. As in the case of most of the developing countries, Indian government too chose centralized management of local level natural resources, vesting it in administrative bodies. In India, at present, majority of forests (more than 80%) are under state control. Eversince nationalization turnef the resource into 'open access', which was once upon a time a 'limited access' common-pool resource, forests that have been closer to human dwellings are prone to abuse. The exclusive ownership and management of the resource by the state has been practiced for more than 150 years and has proved to be fruitless. This can be said because neither the target of bringing 33 per cent of country's total geographical area under forests (as stipulated in subsequent National Forest Policies) has been achieved nor has it been able to maintain the forest cover that existed at the time of independence. On the other hand, the communities (around 400 million people) residing in and around forests do not have any autonomy regarding the use of forests. In strictly legal sense, they cannot make rules of use, they cannot punish anyone for misuse, nor can they take initiative in improving the condition of forests, and create any stakes in it. This is a major factor discouraging communities to invest time and effort in the protection of the resource. Deforestation can be attributed to several factors like growing population pressure, technology and industrialization, commercial logging, poverty, and the policy of the government of exclusive management by bureaucracy etc. The apparent impact is – conversion of forestland for non-forest uses. Forest areas, especially the ones adjoining human habitats, are more likely to be converted into agricultural fields. In Indian context, it is not possible to forget that the communities in pre-British era had location specific informal institutions with elaborate rule-structures ensuring sustainability of the resource (Roy Burman, 1985; Gadgil, 1938; Ghate, 1992) Although the per capita forest availability was large and the resource was not threatened at that time, the then leadership crafted institutions keeping in mind the inevitable eventuality of unrestricted use. Present day situation not only reiterates the need for such institutions, but also their inevitability. The British rule and the thoughtless imitation of their policies that followed in the post-independence era created a situation where the forest-dwelling communities did not have any stake in forests in return of their sustainable, restricted use. The fact that public policies based on the notion that all Common Pool Resource (CPR) appropriators are helpless and must have rules imposed on them, can destroy institutional capital that has been accumulated during years of experience in particular locations (Ostrom, 1990), did not strike the then policy makers. The inevitable tragedy of commons that followed, however, has had varying responses and reactions.

Analysis of the causal factors responsible for the present scenario has brought forth, a now well-accepted fact, that *de facto* land use decisions of the local populations are more determining than the *de jure* official decisions. Although the figures of encroachments on the land under forests are soaring, there also are studies indicating existence of communities that are consciously maintaining and managing the forests within their village boundaries, at their own initiatives. While some village-communities prefer extending their legal boundaries of cultivable land at the cost of forestland, and thus discounting future heavily, there are others who subscribe a very low discount rate at the cost of present income, and prefer sustainable eco-friendly development practices. It is becoming increasingly clear that local communities both filter and ignore the central government's rules. They add their own rules, generating local institutions – rules-in-use – and patterns of activity that can diverge widely from legislators' and bureaucrats' expectations (Gibson, McKean, Ostrom, 2000: 3). A fundamental puzzle that intrigues researchers as well as policy makers is, while some communities organize themselves to solve the problems of institutional supply, why others in similar circumstances do not? In which circumstances are people, who face a potential threat of 'tragedy of commons', able to organize a system that averts it? What accounts for variations in ecologically homogeneous areas? These are some of the questions to which scholars are seeking answers for the last two decades.

From 'Open Access' to 'Common Pool Resources'

CPRs attracted attention especially after Olson's oft-cited statement, "indeed, unless the number of individuals in a group is quite small, or unless there is coercion or some other special devise to make individuals act in their common interest, rational, self-interested individuals will not act to achieve their common or group interests" (Olson, 1965: 2). Obviously, Olson centred his theory of collective action around the concept of 'public good'. The argument was further substantiated by Hardin's famous 'tragedy of commons', in his article in *Science* (1968). "Ruin is the destination towards which all men rush, each pursuing his own interest in a society that believes in the freedom of the commons" (Hardin, 1968). This prompted division of scholars into two schools. One advocating privatization as the 'only' way, while the other promoting 'centralization of authority' as the 'only' way. These arguments continued as the only two alternatives until as recently as 1980s. It is because the conventional theory of common-pool resources assumes that the users are short-term, profit-maximizing actors who possess complete information. The resource is assumed to generate a highly predictable, limited supply of resource in the relevant time period. Users are assumed to be homogeneous in terms of their assets, skills, discount rates and cultural views. In this theory, anyone can enter the resource and appropriate resource units, gain property rights on what they harvest and sell it in open market (Ostrom, 1999). This assumption of open access would reasonably result in over-harvesting. However, this type of contribution to commons literature has travelled a long way since Olson's (1965) 'zero contribution thesis' through Hardin's (1968) 'tragedy of commons', to many scholars taking cognizance of voluntary associations that are creating and enforcing rules that protect natural resources. Literature published in the last two decades provided ample evidence of this. Field research has established today that though temptation to free ride on the provision of collective benefit is a universal problem, self-organized resource governance too has survived in multiple generations (Ostrom, 1990). It is also found that when the users of a common-pool resource organize themselves to devise and enforce some of their own basic rules, they tend to manage local resources more suitably than when rules are externally imposed on them (Tang, 1992; Baland and Platteau, 1996; Wade, 1994).

In case of India and the similar developing countries, where resource dependence is substantial, resource use policies of the governments were unfortunately based on the belief of the scholars who advocated centralization of authority. Gradually the forest dwelling communities were treated as enemies of forests (NAC, Ghate, 1992; Guha). For the local people, although the resource was legally snatched away from them without providing for alternatives, their dependence on natural resources continued to prevail. However, on the one hand forests were treated as a source of revenue for state coffers, the federal governments were unable to restrict resource use by the local population on the other. This resulted in severe degradation of forests. This situation primarily affected local communities to whom neither the fruits of development had reached, nor were they given stakes in maintenance or development of the resource as such. Despite this hostile atmosphere few communities in India and in other places world over, opted for self-governance of resources that were lifeline for them.

Realizing the fact that they themselves are the primary sufferers of forest degradation, many communities in India have taken on themselves to protect forests that are within their village boundaries, by restricting use within the community. Some such attempts have been informal and based on mutual understanding alone, while others have been much more explicit with a formulated rule structure regarding inclusion or exclusion of participants, obligations of participants, appropriation strategies, monitoring and sanctions and conflict resolving mechanisms (Ostrom, 1999). In case of smaller CPRs where individuals communicate and interact with one another on day to day basis, they get to learn whom to trust, what effects their actions will have on each other and on the CPR, and how to organize themselves to gain benefits and avoid harms. When individuals have lived in such situations for substantial periods of time and have developed shared norms and patterns of reciprocity, they possess social capital with which they can build institutional arrangements for resolving CPR dilemmas (Ostrom, 1990).

It would not be out of place to briefly define the concept of common pool resources here. Common pool resources are those for which exclusion of users is difficult, or it is possible only at great cost (McKean, 2000). They are subtractable by consumption and over exploitation leads to their disappearance. Their joint use reduces the availability of benefits derived from the resources for others. Steins and

Edwards (1999: 540) classify management of CPRs under four basic types of regimes: (1) "open access," which refers to a "free for all" situation, where rules regulating access to and allocation of benefits from the resource are absent; (2) "public property," where access rights for the public are held in trust by the crown or the state; (3) "private property," where tradable rights are owned by an individual, household or a company; and (4) "common property," where a set of rules is present to govern access to, allocation of and control over the CPR. Communal management occurs when government grants villagers formal control, and also when local residents exercise *de facto* control in the absence of formal rights. In India, where forests are *de jure* 'public property', in most of the places these have become *de facto* 'open access'. In some exceptional cases forests have in fact become 'common property' as communities are governing the forest use not only of themselves but also of others. This study tries to document some of these attempts from Central India.

Institutions do Matter

Study of 'Institutions' is gaining an important place in development studies. Institutions are considered to be playing a determining role in the rate of growth of nations and structuring the direction of development. In Ayres's (1962) language, Institutions are organization structures or segments of society. It is now being increasingly recognized that the institutions have material, social, religious and political aspects and for understanding human behaviour it is essential to expand the range of relevant variable in any analysis. "Development Economics has from its infancy embraced the study of institutions and its validity of fields rests on the proposition that institutions matter" (Adams, 1993:248). Institutions have always assumed important roles since 1955, when Sir W. Arthur Lewis, while linking economic development with the study of institutional arrangements, had raised questions like: how do beliefs and institutions change? Why do they change in ways favourable to or hostile to growth? Which kinds of institutions are favourable to growth. He was also interested to know if institutions are inimical to efforts, to innovations or to investment; and to know the realm of beliefs, valuations, ideas of the right way to live within an institutional structure. Initially, the critics of development economics, for a long time, were not ready to accept that institutions were of crucial significance, and there happened to be few studies to prove them wrong. Research has come a long way since then, and beginning from Adleman and Moriss (1967), many empirical studies using sophisticated techniques have successfully established the importance of institutions in development. Even orthodox economists have recently begun to think anew about institutions and their capacity to shape individual conduct and channel processes of change and development. It is now firmly believed that "institutions are humanly devised constraints that structure political, economic, and social interaction (North, 1991). In all countries i.e. First as well as Second and Third world, renewed attentions is being given to institutional arrangements as the fount and matrix of economic growth (Adams, 1993: 258). Institutions are important because they are transaction cost minimizing arrangements. They change and evolve with changes in the nature and sources of transaction costs with the objective of minimizing the transaction cost. Institutions resulting from collective action can have effects on the economy, which may constitute an important first link between collective action and development (Nabli and Nugent, 1989: 112)

Institutions are basically rule structures regulating the behaviour of communities. They represent values and culture of the people. Institutions form a unit of regulated social space, within which people operate and are linked within the domains of other institutions. Also, a society's institution moulds its economy, and as each society is different, so is each economy (Adams, 1993). Therefore, to understand how a community conducts itself, it is necessary to break it down into components or arrays of behaviour. Institutions are commonly understood rules and norms framed by the people, helping them in deciding what actions are required, permitted or forbidden in society. Institutions need not be formal, legally recognized, or written, although many times they are. Institutions may be developed by people seeking to regulate their own interactions or originated with external actors e.g. government, religious organizations, aid agencies (Poteete, Ostrom, 2002). McCay (2002) assumes a broader conception of institutions that includes patterned behaviour as well as rules that locate institutions as major features of the cultural, cognitive, and ecological realms within which acting and decision-making individuals and social groups are embedded.

"Demand" side of institutions is thus well established. Social science research in existing institutions as well as their changing structure can help in understanding the "supply" side of institutional change. It may lead to better understanding of self-initiating group or role of leader and their logical justification in bringing institutional change. In what kind of circumstances do people come together? Do they always need to be led by a person? What is the leader's ability to communicate? What are the means used to make his logic understandable to other members of the group? How does she/he convince the fellow members of the benefits being higher than the cost? Answers to such questions enrich the body of literature on institutional studies. The institutional development economics incorporates study of three things: structure, process, and incentives. These are applied for analyzing many problems in the field of developmental economics, including 'collective action' in CPRs. In this book more emphasis is being given on the 'process', adopted by the communities in the five case studies. 'Process' here is taken to mean sequence of socio-economic activities that have shaped the institutional structure. It is also pertinent to mention how a 'community' is defined.

Defining a Community

Contrary to the bureaucratic belief that locals are enemies of forests, a growing awareness among scholars and practitioners that the actions of local people greatly determine the success or failure of schemes regarding natural resource management is heartening. The role of local people at the local level is crucial because national governments rarely possess enough manpower or money to enforce their laws adequately, prompting many officials to consider decentralizing authority over forest resources. Because local communities that live within forests are primary users of forest products and create rules that significantly affect forest condition, their inclusion in forest-management schemes is now considered essential by many researchers and policy makers (Arnold, 1992). This requires defining a 'community'.

"What better definition of community than shared past, present, and future?" (McCay, 2001:185). 'Community' indicates a group of people living in a defined limited space, with a homogeneous social structure despite several internal heterogeneities. It is mainly because of the common norms shared by the members over a considerable period of time. It is measured by the presence, absence, or strength of shared beliefs and preferences; some stability in membership; some expectation of future interactions; and direct and multiple kinds of relationships among members. People who live and/ or work together share a sense of identity and belonging (therefore, some notion of boundaries and membership criteria), where they share some level of dependence on or caring for the resource in question, and where they also share many norms and goals, they are more likely to be able to develop institutions appropriate to deal with challenges they face in using common-pool resources (McCay, 2002: 385).

The literature on institutions for common-pool resources seems to emphasize the role of 'communities'. Of late, the notion, that local communities can and should play a major role in conservation and environmental management, has been the primary belief of important policy makers at international as well as national level. Many international NGOs and academicia too have come to share this belief. The reason is simple. As has been repeatedly experienced all over the world, where the resources such as fodder, fuel-wood, fish and wildlife, that are intrinsic to everyday livelihoods and household budgets are concerned, even well-funded coercive conservation generally fails (Agrawal and Gibson, 2001:5).

In this book we have used the term community synonymously with 'villagers'. As is well known, majority of Indian villages is a combination of various castes and classes. Therefore, community and its relationship to common-pool resource management becomes meaningless without further specification regarding peoples within their environments, their histories, their cultures, as well as regional, national, and global relations of wealth and power (McCay, 2002: 387). The case studies presented in this book bring out the cohesiveness of each village as a 'community'.

Present Study

"A self-governed forest resource is one where actors, who are the major users of the forest, are involved overtime in making and adapting rules within collective choice arenas regarding the inclusion or exclusion of participants, appropriation strategies, obligations of participants, monitoring and sanctioning, and conflict resolution" (Ostrom, 1997). This study presents five case studies that are at different levels of self-governance. We want to strongly argue against the widespread belief that users of natural resources are always helpless when it comes to managing their commons and cannot do anything about resource degradation on their own. The case studies presented here focus around institutions that have evolved or are evolving around forest – a common-pool resource. The attempt is to show that common pool arrangements can result in effective use, equitable allocation, and sustainable conservation. However, I am completely aware of the fact that it could be the contextual factors that have framed these institutions. Similar institutional rules may not have similar outcomes. How effectively other communities would be able to govern their resource with a similar set of rules would depend upon variation in the biophysical, social, economic, and cultural contexts. The study of 'structure' has

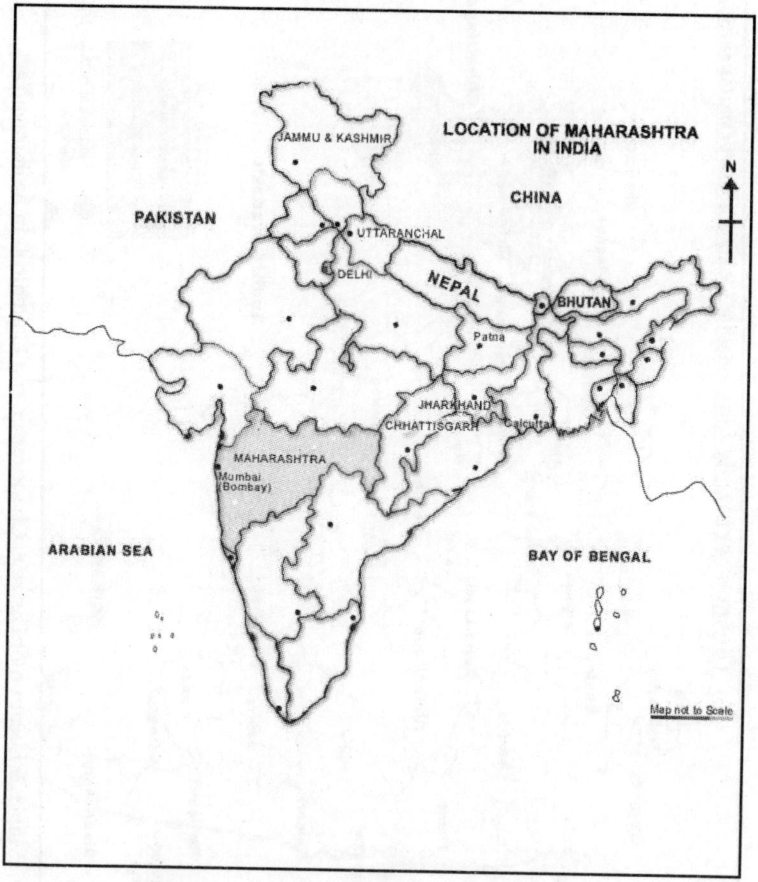

Map 1.1: Location of Maharashtra in India

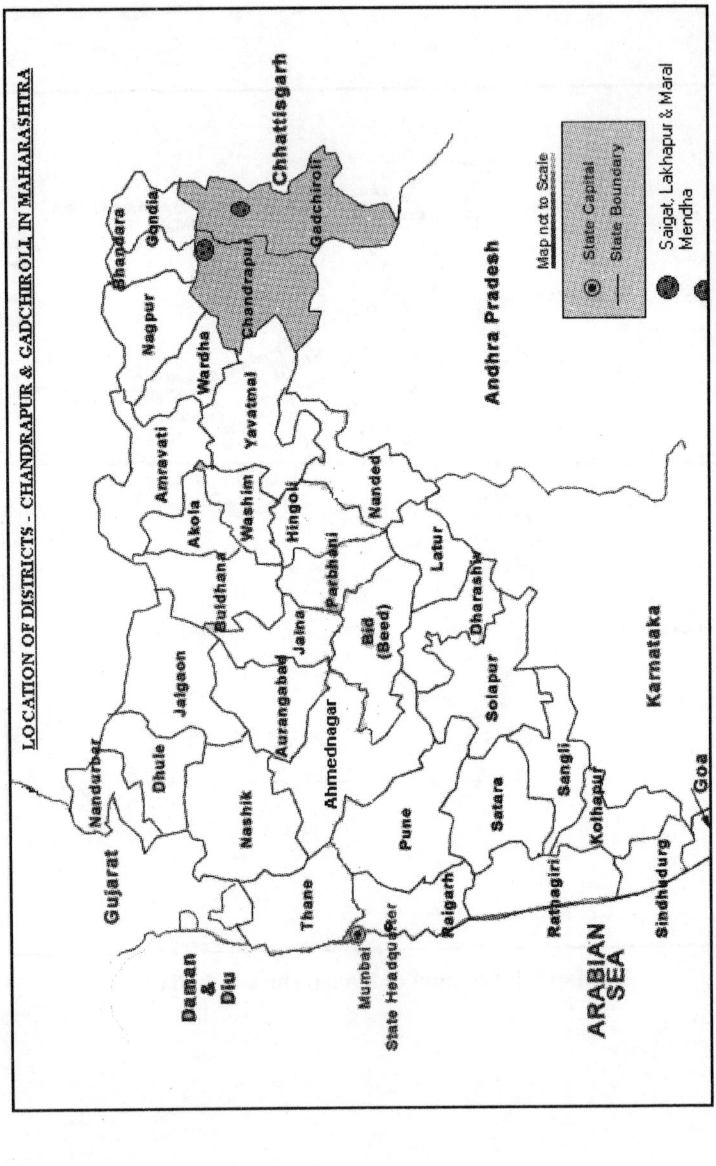

Map 1.2: Location of Districts Chandrapur and Gadchiroli in Maharashtra

helped in comparing these five cases, while the recognition of process has helped in understanding the evolution of the institutions. The evolved system helps one understand how incentives and disincentives regulate individual's conduct and provide means of social control.

We also learn, from this study, that members of some communities take longer to realize growing scarcity of their local forests, and fail to create effective rules to stall over-harvesting, by enforcing their own rules. In this regard local actions vary substantially. Explaining why some communities effectively self-organize while others do not, requires that we understand the benefits and costs of self-organization as perceived by diverse members of the community. It also is affected by the size of the group and the heterogeneity. However there are contradictory claims on both the accounts.

Backdrop of the Case Studies

Five villages that were studied are : Saigata, Maral Mendha, Lakhapur, Lekha, and Mendha. While Saigata, Maral Mendha, and Lakhapur form a cluster of adjoining villages in Brahmapuri sub-district of Chandrapur district, Lekha and Mendha form another in Gadchiroli district (see Maps 1.3 and 1.4). Both the districts i.e. Chandrapur and Gadchiroli, have a high share of total forest area of the state of Maharashtra. In fact Gadchiroli district was carved out from a bigger Chandrapur district in the year 1981.

Each of the five cases presented here, are uncommon in one sense or the other. Since almost 400 million people in India live in villages that are located either within the forest or on its fringes, their dependence on products harvested from these forests is understandably substantial. Yet, lack of ownership or clear-cut rights on forest products make them vulnerable to discretionary treatment at the hands of government officials. Lack of alternatives, poverty, and anti-people attitude of the forest bureaucracy compels the local communities to live like thieves in their own surroundings. Majority of the people consider themselves lucky if they are

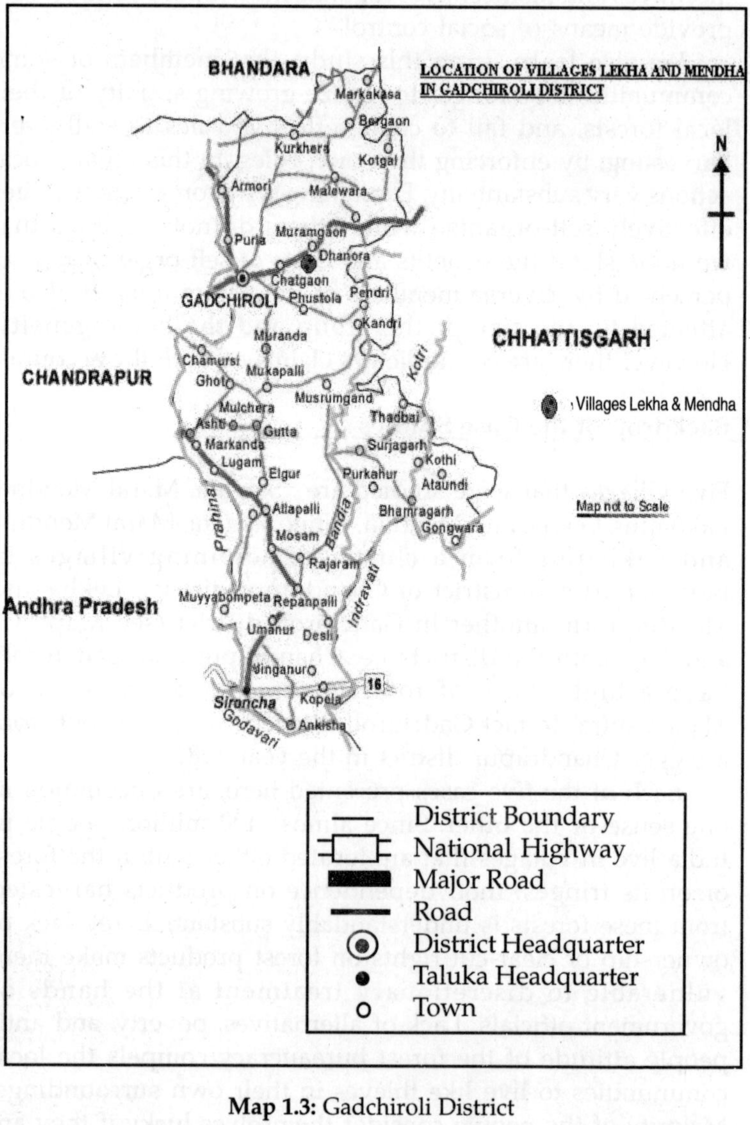

Map 1.3: Gadchiroli District

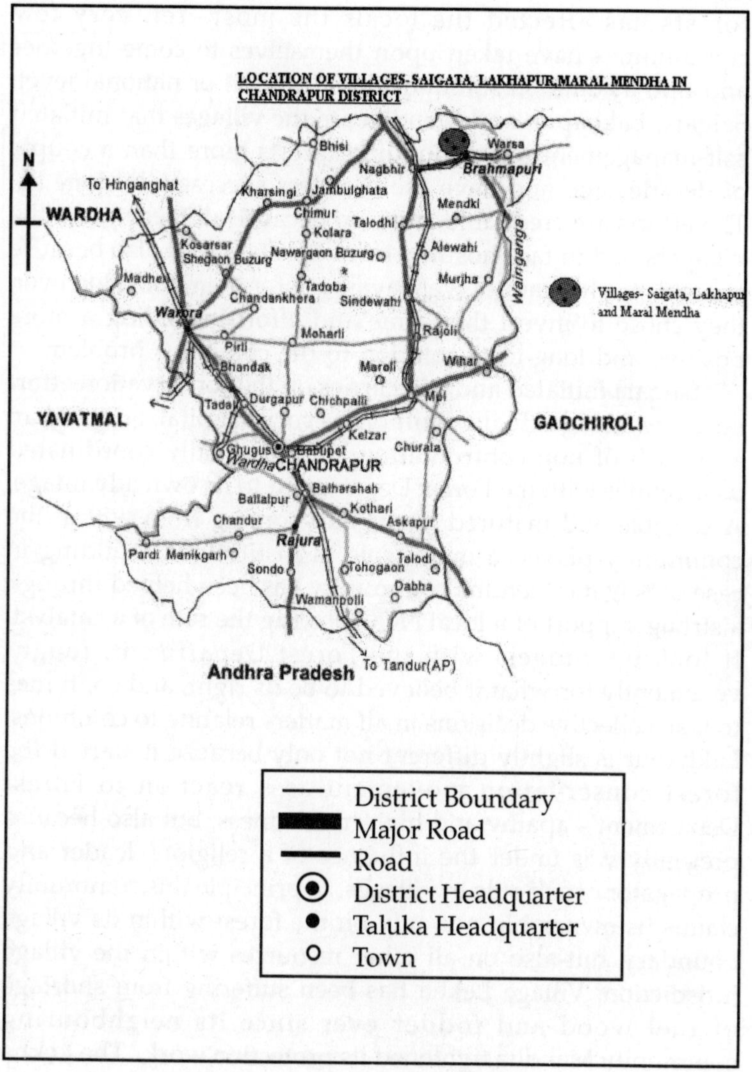

Map 1.4: Chandrapur District

not caught by a forest guard while collecting fuel wood or other small timber from the surrounding forest. If caught, one has to bribe the official to let him go. Deterioration of forests has affected the locals the most. Yet, very few communities have taken upon themselves to come together and find a common solution, either at local or national level. Saigata, Lakhapur, and Mendha are the villages that initiated self-management of surrounding forests more than a couple of decades ago and have more or less successfully done so. The efforts are creditable, more so because all these years the villagers had to face hostility from the state, and also because anarchistic alternatives were available for them too. However, they chose to invest their time and effort in finding a more positive and long-term solution to the commons problem.

Saigata initiated and still carries on the conservation effort on its own. It gradually built on its social capital, adopted an approach of non-confrontation, and gradually coordinated its activities with the Forest Department to its own advantage. A sensible and matured leadership coming from within the community played a major role in institutional building in case of Saigata. Mendha community has been helped through a strong support of a local NGO playing the role of a catalyst. It took up cudgels with the Forest Department, fought vehemently for what it believed to be its right, and continues to take collective decisions in all matters relating to commons. Lakhapur is slightly different not only because it started the forest conservation movement as a reaction to Forest Department's apathy and high-handedness, but also because presently it is under the influence of a religious leader and propagator of self-rule in villages. In principle this community claims its ownership not only on the forest within its village boundary, but also on all other resources within the village jurisdiction. Village Lekha has been suffering from shortage of fuel wood and fodder ever since its neighbouring community Mendha tightened its protection work. The Lekha community has been trying to get together to solve their problem of commons for the last ten years, but it is only recently that they have been able to get into a negotiation with the Mendha community and are now able to put up collective action in cooperation with their neighbour. The case of Maral Mendha community is absolutely different from the other four communities studied here. It was the first among the five case studies (as is claimed by some of the community members) to have started collective action in forest protection way back in 1972. But the effort was abandoned soon as all the households acted independently with immediate self-interest in mind. Majority of the households today, have encroached on forest land, converting it into cultivable land that has resulted in enhanced incomes for the households. Collective action for protecting forests could not sustain here despite the community's fair dependence on forest. They have now become 'free riders' on the forest protection efforts of their neighbours, namely Saigata and Lakhapur. In this way each community has a unique story to reveal.

Roots in History

The cluster of Saigata, Lakhapur and Maral Mendha has an interesting historical background. It would be pertinent to briefly look at post-independence developments in formal institutional changes and the changing attitude/priorities of the Forest Department, to appreciate the actions of the locals. Tracing back the causes of deforestation, it is found that till 1955 large chunks of forests were under the *malgujari* system (this system was introduced by the British by appointing a local rich land holder (Malgujar) to collect revenue on their behalf). The system was abolished in 1955. Before the forests were transferred to the Revenue Department and subsequently to the Forest Department, a large number of trees were cut down by the 'Malgujars' within this short period of transition. Local population also utilized this period to plunder the forest resources. After that the rights were transferred to the Revenue department and then subsequently to the Forest Department. Local communities, however, continued to enjoy some usufruct rights over the forests known as *nistar* rights. But, the Forest Department resorted to its working plans, under which forest harvesting was done through the contractors by auctioning of coupes. Irregularities practiced under this system brought in most of the destruction. Even after the department took over the forestry operations, the attitude of revenue-maximization remained. Personal assessment of forest officers was done, based on the amount of revenue his division could generate. Therefore, although forestry work was done according to Working Plans, developed scientifically, the discretionary power given to officers was always used for maximization of revenue. This is the major reason of deforestation in case of the three villages namely Saigata, Lakhapur, and Maral Mendha in Bramhapuri sub-district, where forestland was already less and population pressure was constantly growing.

The case of Lekha and Mendha is slightly different. Three villages Lekha, Mendha, and Kanhantola were under one *Malgujar* till 1955. The land around Lekha being flat, fertile, and irrigable, being close to a stream, villagers from all the three villages were allotted pieces of land for cultivation here. Hilly, scrubby and infertile forestland on the south of Mendha was kept aside for fulfilling forest-based needs of all the three villages. After the abolition of *malgujari* system, these villages were separated by clearly demarcating the boundaries. Because the cultivated land was in individuals' names, the Mendha and Kanhantola villagers continued to till their land falling under Lekha's jurisdiction, but Lekha and Kanhantola villagers lost their right to use the forest that now fell in Mendha's jurisdiction. As the awareness about sustainable forest use grew because of large-scale deforestation and scarcity of forest products felt by all, Mendha community started consolidating its authority on its forestland by refusing the neighbouring communities to harvest from their forest. Lekha, which had less forest land within its borders, had to struggle for a long time for its day to day forestry needs.

The typical features of the five case study villages are:

Saigata : Local leadership; heterogeneous population; coordination and cooperation with the

		Forest Department; thoughtful and well-planned management of forest; future economic value of forest considered important.
Mendha	:	Traditional/tribal society; homogeneous population; self-declared autonomous ownership of all natural resources within the village; after initial struggle with the Forest Department, now the community cooperates; highly conscientious group.
Lakhapur	:	Disillusioned by Forest Department's inaction and malafide intentions of officials; community considers itself to be the sole owner of all the natural resources in the village; no external aid including that from the Forest Department is received; forest protection is the major activity in forest management; numerically dominant caste on important positions in decision making body.
Lekha	:	Heterogeneous population; weak leadership; lack of initiative; no role of any non-governmental organization; neglected by the Forest Department.
Maral Mendha	:	Internal conflicts of interests; community dominated by one caste; massive encroachment on forestland in connivance with Forest Department officials; immediate income from agriculture considered more important than uncertain long run economic benefits from forest.

Studying these communities is worthwhile because in all the cases anarchistic solutions were available, yet all communities except Maral Mendha preferred sacrificing immediate benefits in order to preserve the resource for future. In situations where population pressure is continuously growing, the demographic impact is mostly viewed as a threat to the resource assuming that it would only lead to degradation. That human beings can also design suitable institutions, for not only protecting but also enhancing a resource, is seldom an assumption. This study urges those interested in effective conservation projects along with successful economic development to base future strategies on the lessons learnt here.

Method Used

All the data required for the study was collected through a set of ten research instruments developed by **'International Forestry and Resources Institutions' (IFRI) Research Programme**, to facilitate collection of information about demographic, economic, and cultural characteristics of communities dependent on forests (Appendix). IFRI also uses rigorous forestry techniques to measure the impact of institutions regulating the use of forest products, on forests. Each research instrument includes variables that are used to understand connection between the physical characteristics of the forests and human interaction with the environment. It is a well accepted fact that the factors that affect the biophysical forest conditions are demographic (population size, rate of change etc.), economic (market demand, shifts in relative prices, size of land holdings), ecological (climatic, soil characteristics, altitude, slope); and institutional (local institutions for resource management, government policies) (Poteete and Ostrom 2002). The set of pre-structured questionnaire helps in making inter-regional comparisons possible. Relying upon rapid appraisal and traditional interview method, the questions guide the fieldwork ensuring collection of all the relevant data.

Collection of forest data through IFRI method has helped us in choosing 'forest health' as an indicator of the success of institutional structure. For this purpose data on forest such as its age, biodiversity, plant composition, human-use etc. was collected through one of the IFRI instruments, the Forest Plot form[1].

Forest Mensuration

Thirty forest plots were selected from each village by stratified random sampling method for forest mensuration[2]. The strata being: areas frequented by the community members, by outsiders, area used for grazing, for timber collection etc. The attempt is to evaluate forest conditions under different community-initiated forest management regimes. IFRI's interdisciplinary methodology not only allows comparisons between actual forest conditions and patterns of use associated with particular institutional arrangements but also contributes to research on collective action required for institutional development and survival. This is important because national policies that enhance the capabilities of local users to create institutions that have legal standing, that provide reliable information about the effects of diverse management strategies, and that back up local monitoring, sanctioning, and conflict-resolving mechanism are important vehicles for improving forest conditions in many countries over time (Poteete and Ostrom, 2002).

Chapter Scheme

After this introductory discussion, experiences of the five communities are presented in the next four chapters, emphasizing on process documentation. Beginning with Saigata, I present the study of Mendha, Lakhapur, Lekha, and Maral Mendha, each case representing a different dimension of collective action (or lack of it). Analysis of forest data is presented in Chapter VI. Finally, a comparative analysis is presented with the help of techniques like multi-criteria analysis and ANOVA followed by theoretical discussion because "until a theoretical explanation – based on human choice – for self organized and self-governed enterprises is fully developed and accepted, major policy decisions will be undertaken with a presumption that individuals cannot organize themselves and always need to be organized by external authorities" (Ostrom, 1992:111-112).

NOTES

1. 'Plots' are demarcated areas for studying the ecology of a forest. IFRI researchers use them to identify trees, saplings, and herbaceous matter found in the forest they are examining. By carefully determining the number and distribution of plots according to the major objective of the study, IFRI researchers can hypothesize how the local population uses (or abuses) forest resources. Plots, in other words, provide the key link between the social and institutional data. The *Forest Plot Form* records names, extent of cover, and the size of plant species within each forest. In this study, the information collected on each plot is aggregated to describe the forest as a whole. Important values of tree species based on density, frequency, and dominance, as well as diversity indices, are calculated from this data. Biodiversity, size class structure, and abundance of plants ranked as priority species by local user groups are also worked out.
2. This method requires placing a grid over a map of the forest and, using a random number table, selecting coordinates of a "random point" in the forest. Using the table of random numbers, plots are determined. The team locates the random points in the forest by using established landmarks such as streams, trails, or large trees that have been marked on the map. If the forest is small, it may be possible to pace off and mark two reference lines crossing the forest perpendicular to each other. Once one plot is found, teams can use compass bearings and paces to reach the other plots. For this study, however, GPS is used for identifying the marked plots. Once the pre-determined plot is reached, centre is marked with a stick or surveying flag, and a boundary of the 1-metre radius circle is also marked. The other two circles are also easily marked off by walking around the centre stake at radial distances of 3 and 10 metres. Once the plot is demarcated, measuring and counting plants starts in the small circle, working outward. Ground cover includes species of grass. In the 3-metre circle the DBH and height of all the saplings are measured, while in the 10-metre circle, DBH and height are measured for all the trees using a Clinometer.

REFERENCES

Adams, J. (1993). "Institutions and Economic Development: Structure Process and Incentive", in Marc Tool (ed.), *Institutional Economics: Theory, Method and Policy*, Kluwer Academic Publishers.

Adelman, I. & C.T. Morris (1967). *"Society Politics and Economic*

Development: A Quantitative Approach", Baltimore: The Johns Hopkins University Press.

Agrawal, A. & Clark Gibson (2001). " The Role of Community in Natural Resource Conservation" in Agrawal & Gibson (eds.), *Communities and Environment,* pp. 1-31

Arnold, J.E.M. (1992). *Community Forestry: Ten Years in Review.* Rome: FAO.

Ayres, Clarence (1962). "The Theory of Economic Progress", 2nd Ed, New York : Schocken Books.

Baland, J. and J. Platteau (1996). *Halting Degradation of Natural Resources: Is There a Role for Rural Communities?* Oxford: Eng: Clarendon Press.

Gadgil, M. (1983). *Forestry With A Social Purpose, Towards a New Forest Policy*, New Delhi: Indian Social Institute, pp. 115.

Gibson, C., McKean and Ostrom (2000). "Explaining Deforestation: The Role of Local Institutions" in Gibson, McKean and Ostrom (eds). *People and Forests—Communities, Institutions, and Governance,* pp. 1-26.

Ghate, R.S. (1992). *Forest Policy and Tribal Development*, New Delhi: Concept Publishing Company.

Guha, R. (1983). "Forestry in British and Post-British India – A Historical Analysis" *Economic and Political Weekly*, Vol. XXVIII, No. 44, pp. 1883.

Hardin, G. (1968)."The Tragedy of Commons", *Science*, 162, pp. 1243-1248.

Lewis. A.W. (1955). *The Theory of Economic Growth*, London: George Allen & Unurim, pp. 11-12.

McCay, B. (2001), "Community and the Commons: Romantic and Other Views", in Agrawal & Gibson (eds). *Communities and Environment,* pp. 180-192

McKean, M. (2000), "Common Property: What is it Good for, and What Makes it Work?" in Gibson, McKean and Ostrom (eds). *People and Forests—Communities, Institutions, and Governance,* pp. 27-56.

Nabli, M.K. and J.B. Nugent (1989), "The new Institutional Economics and Economic Development: An Introduction", in Nabli & Nugent (eds.) *The New Institutional Economics and Development: Theory and Application to Tunisia*, Amsterdam, The Netherlands: Elsvier Science Publishers.

National Commission on Agriculture (1972), Interim Report on Production Forestry- Man made forests, Government of India, New Delhi.

North, D. (1991), "Institutions", *Journal of Economic Perspectives*, 5, pp 97-112.

Olson, M. (1965), *"The Logic of Collective Action: Public Goods and the Theory of Groups"*, Cambridge, M.A. Harvard University Press. pp.2

Ostrom, Elinor (1990) *Governing the Commons: The evolution of Institutions for Collective Action.* New York : Cambridge University Press, pp 182-216

Ostrom, Elinor (1992) "Policy Analysis of Collective Action and Self-Governance" In William N.Dunn and Rita Mae Kelly (eds.) *Advances in Policy Studies since 1950, Policy studies Review Annual,* Vol.10, pp. 81-119

Ostrom, Elinor (1997) "Self Governance and Forest Resources" Conference on 'Local Institutions for forest management: How can Research make a Difference', CIFOR, Bogor, Indonesia, November 19-21

Ostrom, Elinor (1999) "Self Governance and Forest Resources," CIFOR occasional paper no. 20, Indonesia.

Poteete A & Ostrom (2002) " Institutional approach to the study of Forest resources", in John Poulsen, (ed.) *Human impacts on Tropical forest biodiversity and genetic resources,* Centre for International Forestry research. (CIFOR) New York: CABI publishing

Roy Burman, B.K. (1985), "Issues in Environmental Management Centering Forest and Role of Tribal Communities", *South Asian Anthropologist* , pp.43

Steins & Edwards (1999), "Collective Action in Common-Pool Resource Management: The Contribution of a Social Constructive Perspective to existing Theory", *Society and Natural Resources,* 12: 539-557, 1999.

Tang, S.Y (1992), *"Institutions and Collective Action: Self-Governance in Irrigation"*, San Francisco: ICS Press.

Wade, R. (1994) *"Village Republics: Economic Conditions for Collective Action in South India"* , San Francisco: ICS Press.

2

SILENT REVOLUTION IN SAIGATA

Appropriators from CPRs have repeatedly shown their capacity to organize themselves, establish credible commitments, monitor each other's behaviour, and impose sanctions on those who break their commitments.

—Elinor Ostrom

The occasion was that of a wedding. The whole village had gathered to witness the proceedings. The bride was ready and the bridegroom's party had also arrived from the distant village. The atmosphere was charged with anticipation. At about 10 in the morning when the wedding ceremony was still in its initial stages, the community's forest guard came running with the message that the forest adjacent to the village had caught fire and it was fast spreading to the other parts of the forest. This news had such an impact on the gathering that within minutes the *pandal* was empty as everybody including the bride's relatives had left the wedding venue, to put the fire out. The bridegroom's relatives and the bridegroom himself was taken aback to see everybody rushing, but when they saw the bride herself getting rid of her cumbersome ornaments and flowers to join the fire extinguishing party, they did not waste much time in deciding to join them too. After some time, the villagers, assisted by the guests had successfully put the fire out. The interrupted wedding was once again started.

This is a true story of a village called Saigata, which is eight kilometres away from the Brahmapuri sub-district in

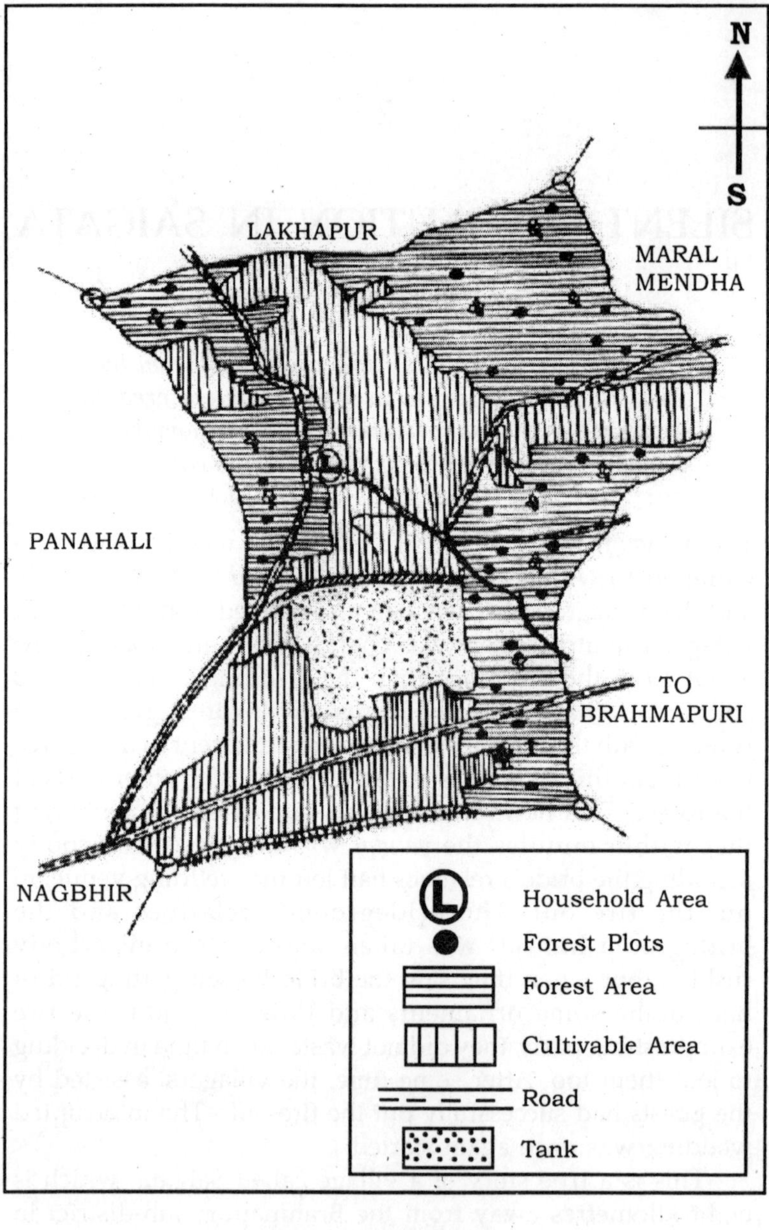

Map 2.1: Village Saigata

Chandrapur district of Maharashtra state. This kind of concern and commitment on the part of a community, which was once responsible for the degraded state of its forest, is certainly an uncommon occurrence. Saigata is an old village, which came into existence somewhere around 1700 AD. Although there are no published or written records about its history, there is a story about how it came into existence. The villagers believe that around three hundred years ago this settlement started taking shape when few cultivators came to this area looking for cultivable land and settled down. However, they gradually migrated to other villages while others kept replacing them till the present settlement formed. Earlier, majority of the people in the community were of Gurav caste (village priests), but now it has changed into a heterogeneous community, slightly dominated by the Mahar caste (earlier considered untouchables), where the Mahar constitute 15 per cent of the population followed by, Gowari (12 per cent), and Gond (12 per cent). Mass migration of the Gurav community has been attributed to a belief prevalent at that time, that Saigata was a village of 'barren' women. It was believed that women in this village could not conceive. This had instigated the people to leave the village one by one. People talk about the times when no family was ready to marry their daughter to a boy from Saigata. The origin of this myth may have been that the wife of the then *malgujar* did not have any children, which was considered ominous.

Recent History: Not So Glorious

Saigata once had a lush green forest with rich floral as well as faunal biodiversity. This was not very long ago, as many people living today narrate their encounter with tigers and other wild life. Saigata village was a corridor connecting two Wildlife sanctuaries – Andhari Sanctuary which surrounds Tadoba Tiger Reserve on its West (around 20 kms) and the recently created Bhamragarh Sanctuary to its south-east (around 100 kms). The villagers do not have to dig deep in their memories to remember how the forest in the village boundary not only provided them with sufficient fuel wood, fodder and timber for household use, but also generated excess – a source of income for the locals.

Area

Total area	566.44 ha
Forest Area	280.00 ha. (49.43%)
Per capita forest	0.65 ha.

However, the unsustainable use by the community, the revenue maximizing approach of the Forest Department, and growing number of settlements in the neighbourhood, nearly wiped the forest out.

The process of degradation had multiple dimensions. All the households in the community were migrants who had settled down not very long ago. Therefore, the community was not a cohesive group with shared past and developed mutual trust. The formal structure of governance was also changing in 1955 from *malgujari* system to forests being transferred to the Revenue department as Protected Forests, where the community had usufruct right known as *nistar* rights. Two decades later, the ownership of forests changed once again, as the Forest Department took over. It also tried to consolidate its hold on the forest with the single purpose of revenue maximization. On the other hand, cultivation being the sole occupation of all the communities in the vicinity, encroachments on forestland were also not uncommon. Growing population pressure, small size of land holdings, and meagre incomes of the landless, all resulted in unsustainable use of the resource. Forests surrounding the villages being 'public property', the community living at subsistence level, looked at it as a source of cash income. The upcoming township in Bramhapuri, that was declared a sub-district centre, was only five kilometres away from Saigata as well as Lakhapur and Maral Mendha. This fact accentuated the speed of forest degradation in the area. Lot of felling took place in Saigata forest for supplying timber for constructions in Bramhapuri. Increasing number of houses and households meant increasing demand for cooking fuel, bamboo, and timber for furniture etc. For all this nearby forests was an easy and cheap source. Principle of 'profit maximization' prevailed on the 'suppliers' of forest produce in the wake of growing 'demand'.

Extending Collective Effort

One Mr. Suryabhan Khobragade, a resident of Saigata, was always uncomfortable with the constant degradation of the forest. He was disturbed by the fact that a part of the forest, which was once known as *andhari*, meaning darkness only because it had such thick tree cover that even sunrays did not reach the earth below, was now totally devoid of big trees. He had witnessed the changes in land-use patterns since the days of the *malgujar*, working for him as child labour. Between 1955 and 1975, the forest around Saigata had changed from thick canopy forest to degraded land. Yet, he was also aware of the fact that it would be difficult to dissuade the fellow villagers from giving up their income generating pursuits without offering them an alternative. After many discussions with the like-minded people in the village, it became clear that asking the fellow villagers straightforwardly to stop anti-forest activities would not yield the required response. Instead, something positive needed to be done circuitously, to bring the community together. It had to be an activity that gave positive returns to encourage people to come together in the first place. At that time, Agricultural department of the state government had a scheme for training farmers under *Krishak Prashikshan* for adopting improved techniques of cultivation. Under the scheme, there was a provision of setting up a *Krishak Charcha Mandal* (Farmers Discussion Forums) in any village that showed interest. Mr. Khobragade had attended a five days training workshop under the scheme where he was given a radio to listen to the agriculture related programmes. He came back and spoke to villagers about his experience and the things he learnt at the training programme. They all then got together, and on 7th October 1973, as

Mr. Khobragade distinctly remembers, set up a *Krishak Charcha Mandal* in Saigata.

At that time a radio programme discussing various agricultural techniques from all over the world was broadcast by All India Radio (AIR) on Mondays and Fridays, wherein the farmers were encouraged to apply these techniques. All the men of Saigata community religiously got together at 7:30 in the evenings of those days to listen to the programme. The programme was of half an hour duration but the Saigata community used to continue their discussion over the techniques given in the programme, for another half an hour. This was the only occasion at that time when the people of the village came together. As a part of the scheme an award at *Zilla Parishad* level was announced by the Agricultural Department of the Chandrapur district for those communities who successfully applied the techniques taught through the radio programme. They were given a questionnaire for this, made available at the sub-district office of AIR. A group of inspectors from the Agricultural Department used to visit the village to confirm the actual application of the technique and evaluate the genuineness of the effort. The farmers of Saigata started successfully applying these techniques. This got them awards for two consecutive years in 1974 and 1975. It proved to be a great morale booster for the villagers as their very first experiment in collective action had brought them some recognition. The economic benefits that followed this successful experiment, together with recognition of collective effort in the form of the award helped the community to get together for reasons beyond farming.

Minor Cause of a Major Change

Mr. Khobragade's discomfort with the degraded state of the forest turned into a crusade to rescue it from the brink of destruction. It came about when one morning he woke up to find his mother warming bathing water for him over cow-dung cakes (often used as fuel in rural India). This brought to his mind the fact that the 300 hectares of abundant forest that once belonged to Saigata, not in a very distant past, has reached this state of degradation. Those were the days when there was enough fuel wood available in the forest, and all the cow dung was used more efficiently as farm manure. He shared his anguish with the community. The community had by this time already come together through the *Krishak Charcha Mandal* and a majority of them shared his concern of the deteriorating condition of the forest leading to scarcity of fuel wood and fodder. All that they needed now was a plan to start the process to save the forest. Mr. Khobragade initiated an effort to first identify the dependence of each household on forests.

Dissuading the Expropriators

Under the leadership of Mr. Khobragade, the active group within the community found out that there were three types of activities that were seriously degrading the remaining forests. The first category consisted of community members who were collecting fuel wood for sale in Brahmapuri, delivering it literally on the doorstep of the households. They were gradually made to realize that they were actually losing out by doing so, since they were either not charging for the labour involved in collection of fuel wood or for transporting it to Brahmapuri. Mr. Khobragade suggested that either the customers should come to Saigata and buy fuel wood or the collectors should charge an extra price for transporting the head loads. This argument convinced the traders that they were making a net loss in the trade and that they would make a fool of themselves if they continued to do so. This argument proved to be a convincing enough. Gradually the sale of fuel wood by the community members stopped. The second category of extractors was that of charcoal makers. This was a very profitable business, as it fetched more income than farming. Thus, it was difficult to convince this category of extractors without offering any immediate alternative. The pro-conservation group tried to reason with the charcoal makers and convinced them that their business would last only till the forest lasted. After that, they would have no source of income. They were also enlightened on the other ill effects of forest degradation. Third category consisted of all the households of Saigata, who were indiscriminately exploiting forest products, not caring for sustainability of the resource. Everyone was experiencing the lowering of water level in the wells and in the village lake. Pests on crops had increased, severely affecting agricultural production and income. Women had to spend more time and energy in collecting fuel wood and fodder. Bamboo and other useful timber had vanished from the forest due to which people had now started buying it from the market. Therefore, the community had only to convince itself for imposing restrictions on self-consumption. To encourage this, Mr. Khobragade adopted a unique and indigenous method. He used his skills in writing, directing and acting in skits (known as *dandaar* in rural areas of this region) to spread the message. He feels proud while narrating that high caste actors in the skit had to ask him to apply make-up to their faces, thus letting him touch them. The typical style of this folk art of taking a dig at the social ills in a lighter vein, made people think more seriously about future outcomes of their present actions.

Informal Organization comes up Formally

All these factors combined together led the people to protect the forest by setting up *Van Sanrakshan Samiti*, meaning 'forest protection committee' in 1979. The common feeling was "if we do nothing to protect our own forests, who will?" The general body of the committee consisted of all the adults in the village as members, and an executive body was formed to regulate day-to-day functioning. However, the *Gram sabha* i.e. the general body, made all the rules regarding use of forest products etc. Suggestions were invited from all members. As a result, all the members participated in rule making, and decision taking activity. Every decision was taken unanimously, and therefore, was binding on everybody. Initially the *Samiti* started supervising and controlling various activities of its own community members inside the forest. There were only few infractions. However, it was increasingly realized that without stopping the entry of outsiders inside the forest, it would not be possible to protect the resource in the true sense. With this began a long period of struggle for the people of Saigata.

The forest association

Name of the Forest Association	Van Sanrakshan Samiti
Activities carried out by the members of the Forest Association	Protection, enforcement, plantation
Formation of the association initiated by—	Mr. Suryabhan Khobragade
Year of formation	1979
Year in which legal status received	1993
Source of financial support for the association	Voluntary contributions and fines

The revenue boundaries of villages were not clearly demarcated. As a result, movements of people from nearby villages into other community forests, was a common occurrence. Especially forests being *de facto* an open access resource, restricting anyone from illicit extraction was unheard of, as well as unacceptable. Kirmiti Mendha, a neighbouring village, did not have much forest within its village boundary. For all these years, it had met its forest needs from Saigata forest. The Kirmiti Mendha community members did not appreciate the newly imposed restrictions by the Saigata community. The demand for fuel wood continued to rise in Bramhapuri, and gradually even other nearby villages had started experiencing the dearth of forest products due to continued denudation in the region. Despite lot of pressure on the community members, Saigata continued to protect its forest by keeping vigilance day and night. The whole community shared the responsibility of watching the forest. Households took turns in patrolling. Many instances are narrated by Saigata villagers today about how women had to come forth to stop the timber thieves of the same gender and physically seize their axes. This had to be done by the women of Saigata because a woman, if caught, can easily lodge a police complaint on the charge of harassment even if she is merely touched by a man.

Going Beyond Protection

Uniqueness of Saigata lies in the fact that it did not stop at merely protecting the forest. It gradually moved towards thinking in terms of improving the quality of their forest. The people had already come together for practicing improved methods of cultivation. This had helped reduce forest dependence to some extent. Rules and restrictions on the harvesting of forest products were already established. But due to paucity of funds few maintenance activities could be taken up initially. More emphasis was on regulated use and strict guarding of forest from outsiders. As a result, the forest growth till then was only due to natural regeneration. Now, the community started thinking of managing the forest in addition to protecting it. After several brainstorming meetings, it was decided that different patches of forest, traditionally known by different names like *Tiwas, Kotwal, Mothya Devache Raan, Kosran* and *Gavtache beed,* would be treated as management zones by the villagers henceforth. Thus, each management zone was closed for harvesting for a certain period in order to allow the natural regeneration to take place without any disturbance. *Gawatache Beed* (the forest of grass) has been kept separately for the fodder and thatching needs of the community. Similarly, *Andhari Van* (The dark forest) is completely prohibited for any fresh wood cutting activity. Only dead, dried wood is allowed to be collected for the purpose of fuel wood. All these years of protection and plantation are perceivable as the forest is slowly regaining its lost characteristic of darkness (*Andhar*). In Kotwal Raan (*kotwal* is the lowest government official in revenue department) restrictions are on cutting of trees. Only restricted felling is allowed. *Mothya devache van* (the forest of the main deity) and *Kosaran* (indicating a short distance from the settlement) are used alternately by the community for extraction of forest products. *Tiwas Van* (part of forest dominated by *Tiwas* trees) is kept aside for *in situ* plantation for converting it into a mixed forest. It was closed for three years in order to allow the growth of valuable species used for timber, like *Bija (Pterocareus marsupium)*. This was mainly done after being challenged by the Forest Department that *Bija* never grows naturally. The villagers' argument was that by stopping the entry of animals in that area, and thereby preventing the crushing of *bija* seeds under their hoofs, the seeds could germinate. Rightly so, within three years *Tiwas Van* had many *bija* saplings. This indigenous knowledge was reportedly appreciated by the then forest officer.

Fighting Smaller Battles

Changing of forest ownership from revenue department to forest department posed some minor problems for the community. While the revenue department had recorded and accepted the usufruct *nistar* rights of the locals, forest department did not respect these and had its own rules regarding use of forest products by the local communities. The dissatisfaction amongst the communities was accentuated by the fact that the Forest Department was passive when it came to protection and developmental activities, but was trying to assert its authority when it came to use of forest products even for self-consumption. An anecdote is related by the villagers in this connection. *Van Sanrakshan Samiti* (VSS) in Saigata had developed a highly equitable distribution system for grass from *Gavtache Beed*, the zone earmarked and developed exclusively for grass for both thatching and fodder for animals, wherein the landless households spent labour in cutting the grass and selling it in small bundles at a very low price to anyone who needed it within the village. In 1985, Forest Department seized 60,000 bundles of fodder, weighing one kg each, from the people of Saigata and demanded that the people buy it at the rate fixed by the Forest Department, claiming that the community did not have a right to harvest from the forest. The people protested as they already had the *Nistar* permit obtained in 1976, when forests came under the Revenue department. Concerned officer was contacted, but since it was no longer under his jurisdiction, the matter was referred back to the Forest Department. VSS members then met the Range Forest Officer who demanded Rs. 8 per cartload of the seized grass, which was lower than the price at which the landless of the village were selling. The community readily agreed to pay and released their grass.

Inclusion in the Formal Participatory Structure

In 1992 Forest Department asked its divisions all over the state to select two villages each for implementing Joint Forest Management (JFM) programme. Seeing the good work done by the people of Saigata, the Bramhapuri Division of the Forest Department approached them to register under JFM. But the people did not accept the invitation immediately as they did not trust the department's officials. Long discussions took place on questions like, why has the department suddenly changed its stance, what is the guarantee that the revenue will be shared by the department 20 years from now, what is the need for getting into the agreement since the community was already enjoying *de facto* usufruct rights etc. The overall impression that the people had about the Forest Department was not very good due to past experiences. They even blamed the Department for the degradation of the forest. The department had adopted a faulty policy when the Bramhapuri town was being built and as a result the forest was exploited more than ever. Other than that, they had had a few tiffs with the Forest Department over the harvesting of forest products from the Saigata Forest as has been already mentioned. All these incidents caused the initial hesitation to cooperate, on the part of the people of Saigata when the Forest Department approached them in 1992 to set up a Forest Protection Committee under JFM. For six months, exhaustive discussions took place between the people and the Forest Department. The Department was, initially interested in entrusting this work of convincing people to join JFM, to Mr. Khobragade, but he flatly refused and asked the Department to settle the matter with the people directly, saying that decision will be taken only in the *gram sabha*. Finally, on 27th June 1993 this group formally organized itself as the "Forest Protection Committee" under JFM. However, from the records maintained by the department, the year of registration of Committee is shown as 1996. The community believes that this has been done to postpone sharing of benefits with the people of Saigata, by 3 years.

Present Structure of the Association

Despite joining JFM, Saigata community continues to have its own institutional structure for all practical purposes. At present, the forest association has an executive committee, consisting of three female and six male members. The female members have been a part of the executive committee for the last five years, but none of them has ever become the leader of the executive committee. The members of the association hold elections regularly to elect members of the executive committee. The executive committee members meet at least once a month. There is no specific tenure for President, Vice President, Secretary and the treasurer. Their term depends on the number of votes given by the members. Since the inception of the association, three different individuals have led this forest association; they are Mr. Suryabhan Khobragade, Mr. Vishwanath Gaikwad and Mr. Kishan Gaikwad. Those who have held positions in the executive committee have been forest users too. There has not been any competition for any position in the executive committee over the past decade. The committee has always chosen its leader unanimously. The members alone hold the right to remove any member of the executive committee, and no external or higher-level authorities have the power to do so.

All the members of the Committee have been working on voluntary basis since the inception of the association, and nobody receives any remuneration either in the form of money or in kind. None of the members of the executive committee own any capital assets that requires raw material from within the forest for its operation. Most of the members of the executive committee live within the distance of one kilometre from forest. The education level of the members of the executive committee is in tune with the average education level of the whole village. Ethnic and religious diversities exist among the members of the executive committee as they exist in the Saigata community. Religions followed are mainly Hinduism and Buddhism.

Products Harvested

After the community came under the wings of JFM, the user group took up the regeneration activity of plantation, which could not be taken up earlier due to lack of funds. Now it is financed through the JFM funds. This is especially important as the Saigata community continues to depend heavily on the forest for its needs like fodder, fuel wood, timber, food etc. hundred per cent of its fodder, fuel wood and timber (for house construction) needs are met by the community's forest.

Forest use and perception

Products harvested from the forest	Fuel wood, fodder, bamboo, timber for house construction, food supplements etc.
Total number of animals for dependant fodder	1246 (Cattle-160, Buffaloes-60, Goats-1000, Poultry-20, other animals-6)
Increase in the number of households in past 20 years	23
Villagers consider the forest	A sacred as well as an economic resource
Villagers rank their forest as	Somewhat abundant
Techniques adopted to reduce resource dependence	Bio-gas plants installed, improved techniques of cultivation adopted

Even ten per cent of their food needs are fulfilled by the minor forest products like *awala, char,* and *moha*. While the rates of forest products available in the settlement area are low (Rs. 20 for one bundle of fuel wood, Rs.10 per bundle of fodder and Rs. 100 per pole of timber), prices are much higher in the market nearest to the settlement. The price of one bundle of fuel wood (where a bundle if converted in standard metric units of kilograms, comes to 30 kilograms) is Rs. 30, for fodder it is Rs. 20 per bundle (this when converted comes to 20 kilograms) and for timber it is around Rs. 120 per pole (where one pole is approximately 12 kilograms). Thus, plantations like the 215 ha. bamboo plantation carried out under the JFM scheme, are important for the community. Mainly because the close substitutes found for fuel wood, fodder and timber like kerosene, livestock feed, and bricks for house construction, available in the closest market are expensive, the community cannot really afford these. Under the Joint Forest Management programme, the community has legal right to take a share of forest products for self-use and also a share in net profit from the sale of forest products as and when these are harvested.

Giving Teeth to the Association

The whole community of Saigata strives for the success of the association. For this the members of the forest association have carried out many activities over the years. Since 1996 the community has coordinated and regulated following activities:

(a) Plantation of seedlings
(b) Maintenance of the forest
(c) Harvesting of forest products
(d) Distribution of these products among users
(e) Sale of the harvested products
(f) Distribution of revenue from the sale of the products
(g) Determining the time of the harvest of the forest products

(h) Determining the quantity of the forest products to be harvested
(i) Determining the type of technology to be used to harvest the forest products
(j) Determining who is authorized to harvest these forest products
(k) Determining the use that these products can be put to including religious use
(l) Sale of the rights to harvest forest products that users can trade with others
(m) Renting of non-transferable rights to harvest forest products
(n) Monitoring the condition of the forest
(o) Monitoring conformance to rules
(p) Imposing sanctions on rule breakers
(q) Act as arbitrators for the disputes among local users
(r) Interaction with higher authorities

For doing all this, forest association frames and modifies the rules and regulations continuously for forest produce on the basis of suggestions given by the community members. It is one of the important features of the association. The guarding of the forest, which was initially done voluntarily by the villagers, is now being done by four paid guards. Payment is done from the association's coffer to which the

Local indicators of economic state

Local definition of wealth	Enough food, big family, money, existence of natural resources in the village, possessing intelligence.
Local definition of poverty	Those working on daily wages, can't afford a square meal a day, those dependant on others for their living and lazy people.
Existence of economic disparity based on the local definition	Not much

members of the community contribute. The community has devised the structure of contribution, keeping in view the economic condition of each member. Rs. 10 per month is charged from the landless, Rs. 20 from those who own less than five hectares of land and Rs. 30 from those who have land more than 5 hectares.

Other than this, during the last few years, the members of the user group have taken up many other activities like building up a nursery to distribute seedlings for plantation, remove leaves or needle litter from the forest floor, dig fire-line etc. They have even sought help from external authorities to improve vegetation growth. These activities get reflected in the higher density and frequency of saplings and seedlings in the forest.

Ensuring Acceptability

The Association has recorded its mission and objective as 'achieving self-sufficiency'. Emphasizing that the villagers should be able to earn their living in the village itself reflects collective wisdom of the user group. The forest association has created the rules on its own without following the set of rules provided by the Forest department. Members of the user group, together with the officials of the association, enforce the rules. The association believes in cooperating with other associations especially the Forest Department, and has also signed a Memorandum of Understanding to get help from the department. Nevertheless, they create and enforce rules independent of the department, and do not like interference from the department on such matters. None of the rules created by the association has ever harmed the members in any way. Nor has any resolutions been passed that would cut off a particular individual from obtaining benefits from the forest. The association tries to see that the economically deprived individuals get some support out of its activities. As a result, now the landless are working as agricultural labourers, while earlier they did not have opportunities to earn adequately, and were involved in cutting trees illegally. Even the gender disparity has reduced, and now more women are participating in the decision-making process.

There is a well-developed rule structure regarding forest products. Rules exist regarding harvesting, processing or selling of fuel wood, fodder and timber, which affect the harvesting level or use of these products. There also are rules regarding timing and purpose of harvesting. The primary purpose of forming these rules is to ensure sustainable use of the forest i.e. forest products are harvested only for self-consumption. For this the required quantity is determined according to the need of each household based on the number of members in the household. These rules have not changed much over the past 10 years. Other than the restrictions or rules on the quantity harvested of these products, there also are other restrictions, like the specific location within the forest where these products can be harvested. This restriction has changed over the past 10 years according to the availability. Restrictions also exist on the 'form of the product' that can be harvested, for example in case of fuel wood, the harvesting is limited to dead wood only, and in case of fodder only mature grass can be collected. There also are rules regarding type of transport to be used to remove the harvested product from the forest. Either the members carry the harvested product physically or use bullock carts. These restrictions have remained the same over the past ten years. The members of the user group follow these restrictions almost always.

Dealing with Infractions

Regarding infractions, i.e. breaking of the rules, it was found that sometimes infractions took place regarding harvesting rules. People would not follow the restrictions about specified locations. These infractions have more often taken place during the day time and mainly in the summer season in case of fuel wood and timber. For fodder, it has been noticed that, infractions take place in the winter season. The penalty structure is such that it progressively increases with each repetition of infraction. When a harvesting rule is broken for the first time, a fine, which is greater than one day's work but not more than one week's work, is imposed. If the offender breaks the rule for the second time, then the community has decided on a fine, which is greater than a week's work. If the rule is broken for the third time then the offender is made to apologize publicly and his case is reported to the Forest Department. This decision on penalties is taken through a vote of the executive committee. The fine once imposed, is then collected by an official of the user group or the forest association. Most of the users in Saigata comply with these penalties once imposed on them. The local forestry association records these penalties. All members of the user group are allowed to see these records. If a situation arises when the offender refuses to pay the fine, the committee members of the user group expose the offender to public rebuke, and a final request to comply with the penalty imposed, is made. If in an extreme case an offender loses his harvesting right to the three products the same can be restored only through public apology. So far, the community itself has been successful in imposing the penalties and the users have complied. As a result, no external government official had to be called in to enforce the penalty. Government officials were called only once in the year 1998, when some persons belonging to *Kirmiti Mendha*, a neighbouring village, were caught for illicit felling of trees.

Financial Sources

Main source of revenue is the amount of money collected by the association through the member households, fines, and voluntary contributions received from external agencies. Some households also contribute in the form of labour hours, but no elaborate records are maintained. The major item of expenditure is the amount spent for guarding the forest. All the members of the association together make the decisions relating to expenses on various items. The members of these user groups take the decisions without any kind of interference from external authorities. All financial records maintained are available for the members of the community to see.

Transparency, Flexibility, and Participation

There is complete transparency regarding the rules, functioning of the association etc. as all members actively take part in all the activities of the association by attending meetings regularly. All the members of the associations are well aware of the rules, as they are easy to understand and clear in indicating punishment for breaking rules. At the same time, some degree of flexibility has also been introduced in the rules for dealing with the situations of emergency or for handling unusual problem of any member of the user group. Therefore all the members find these rules fair and legitimate. The decisions are taken in the general body meetings that are held at least once every year, and its implementation is ensured by the executive body, which meets more frequently. There is a provision to call emergency or special meetings to discuss immediate problems faced by the villagers. Such types of meetings have been called during the last five years, and the executive committees has made some changes in the rules of the association by considering the suggestions made by some of the members. Few years back some members had suggested that instead of allowing people to collect firewood whenever they wanted any time of the year, there should be restriction on time period when firewood can be collected. Based on this suggestion the fuel-wood harvesting rule was modified. In this manner the annual general body meeting of the association normally frames the rules or modifies them, decides on the location of harvesting etc. On the other hand the Executive committee normally takes decisions on types of penalties to be imposed.

Importance of Leadership

One wonders how Saigata would have been without Mr. Khobragade. Although, the Forest Department legally owns the forest, the community of Saigata has protected it well for the past 22 years. The effectiveness and versatility of the local leadership is beyond doubt, the most significant factor responsible for the success of this participatory experiment. His success in initiating collective action can be ascribed to variety of reasons. Firstly, since he belonged to the community, and thus was sharing all the problems that were faced by the community. His perception of the problem was clear, his understanding of the community was fairly good. He could convince the community of the rationale behind starting the protection activities. Secondly, Mr. Khobragade won the trust of the community due to his sincerity of purpose and transparency in all he did. He always shared vital information with the fellow villagers, sharing benefits of new techniques. "The leader has to actively participate in all that he asks the others to do. He cannot make people cooperate with him by just giving orders", is how Khobragade feels and also acts. He never assumed role of a leader. In fact, he encouraged the others to come forth and sort problems of the community. He narrates an incident of a time when a few households had actively got involved in brewing liquor inside the village. Others in the community were not happy with it as the youth of the village were getting hooked to this drink. They approached Mr. Khobragade to get this activity stopped. He refused and asked these people to come up with a solution themselves to tackle this problem, encouraging them thus, to search for solutions instead of giving ready solutions.

Thirdly, the means he adopted for conveying his concerns were indigenous and very popular. The traditional art form *dandaar*, as has been mentioned earlier, uses pun in its songs and narrations. The music, acting, jokes, etc. are highly enjoyed by the people. Therefore, instead of giving discourses or sermons, Mr. Khobragade chose this particular medium to bring forth stark reality of forest degradation. This made him popular within his community as well as outside. His growing importance amongst the neighbouring communities earned him respect and acceptability within his community, despite his low caste. He still remembers, how the households from upper castes allowed him to sit and eat with them, but not others of his caste. He successfully broke this barrier as well, using the same medium of popular art form.

He succeeded in instilling sense of belonging and ownership of the resource in the community without invoking antagonism against the state. He encouraged cooperation with the Forest Department, which helped in improving the quality of the Saigata forest. From 1993 (when the village formally came under JFM) onwards, the coordination has only improved. The types of activities undertaken by the Forest Protection committee with silent backing of the Forest Department include plantation and maintenance, forest protection and harvesting, decision-making and rule enforcement. Each decision is taken by the Saigata community as a whole, in *gram sabha*. Participatory decision-making is another important factor that has ensured high degree of compliance.

To initiate and maintain collective action in a socially heterogeneous society that has not shared a common past and does not have tradition of collective decision-making, is a challenging task. This was achieved in Saigata, where people were slowly brought together to trust each other in order to achieve a common goal i.e. the revival of the forest. Regular interaction through meetings to thrash out problems and internal conflicts, encouraging everyone to participate and giving opportunity to all to take decisions for the benefit of the community, thus invoking a sense of responsibility and shared understanding helped in building social capital that did not exist in the community. This is the crux of success of the Saigata community in consistently moving towards a common goal.

Basic Details about the Village

Elevation	246 metres ASL (Above Sea Level)
Latitude	20°36′53.2″ N
Longitude	79°47′07.4″ EO
Nearest Town/Market/ Administrative Centre	8 km
Total Population	432
Sex composition	Males (50.69%), Females (49.31%)
Current number of households	88
Nuclear families	88
Literacy	94.91%
Dominant religions	1. Hindu (77.31%) 2. Bouddha (22.69%)
Dominant caste groups	1. Mahar (15%) 2. Gowari (12%) 3. Gond (12%)

3

CONSERVATION WITH DEFIANCE IN LAKHAPUR

"The struggle for fair access to natural resources by people living off these resources, like aborigines and other rural disadvantaged communities is often aimed against state and its coercive instruments of governance".

—Dr. Anil Gupta

It is not very common for researchers to hear a flat refusal for cooperation when they approach a community, explain their intentions and objective of their study. Yet, this is exactly what happened when we first visited Lakhapur, the next-door neighbour of Saigata. A few community members, with whom we had our first encounter, quite unceremoniously asked us to leave the village. They did not find any utility of our study to them and even suspected us of being connected with the Forest Department. The message written on a board at the entrance of the village, that 'welcomes' all visitors to the village, says in *Marathi — "Equality and brotherhood — this is Lakhapur village. Our village is our God. Ask his permission humbly and then enter our village. We rule our village. Water, forest, and land are ours and will remain ours. This is our motto. No Forest official is allowed to enter the village. No one can enter the village without Gram Sabha's permission"*. This is indicative of the sentiments of the people of Lakhapur towards government officials, especially forest officials. Our determination to study this unique village made us approach a local saint, Tukaram

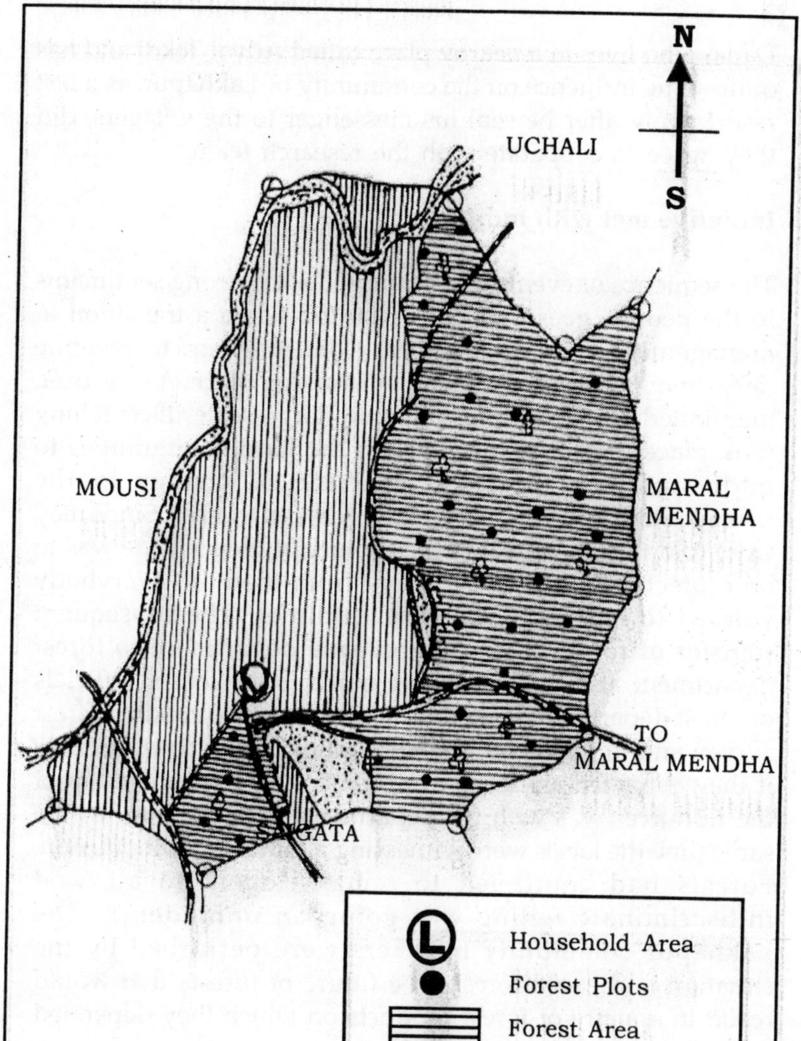

Map 3.1 : Village Lakhapur

Dada, who lives in a nearby place called Adyal Tekdi and has quite some influence on the community of Lakhapur, as a last resort. Only after he sent his messenger to the villagers, did they agree to cooperate with the research team.

Initiative met with Indifference

The sequence of events that prompted such strong sentiments in the people, goes back to the period when a transition in management system of forests, from *malgujari* to revenue department of the government, was taking place. As has been mentioned earlier, during this time large scale illicit felling took place. It took some time for the local communities to understand the practical meaning of change in the management system and its consequences thereof. Since they were told that the forest in the vicinity of their village was to be protected by the government from then on, everybody worked to maximize individual utilities. The subsequent transfer of forest land from revenue department to forest department also did not matter much to the locals. Officials of forest department visited each village only to conduct forest survey and to threaten the communities of dire consequences if they extracted anything from it, because the ownership of the resources was with the department from then on. At the same time the locals were witnessing a different phenomenon. Forests had continued to remain 'open' to all, and indiscriminate felling was going on unhindered. The Lakhapur community members were perturbed by the situation and could foresee the future of forests that would result in scarcity of forest products on which they depended so heavily.

In order to stop this, the community approached the Forest department for their help. Instead of support, the community received a message from the then Range Forest Officer to "mind their own business"! This led to souring of relationships between the community and the Forest Department forever.

Area

Total area	*491.50 ha.*
Forest Area	269.86 ha. (54.91%)
Per capita forest	0.49 ha.

Leadership from 'Inside'

The resource had already degraded to a great extent by the middle of 1970s. One individual Mr. Baburao Dighore from Lakhapur took it upon himself to bring the people together to make them aware of their rights on the forest products. The effort started informally, people usually assembled in the evenings and discussed the day-to-day happenings. Each household's dependence in the forest being substantial, it formed the centre of discussion. Growing concern over the continuous deterioration of the forest quality and unobstructed felling by people from nearby villages compelled the community to think about protective as well as restrictive measures. People, after that, realized that if they do not protect the forest at this stage of degradation, they would never be able to save it and the source of their daily needs would be permanently gone. Thus, they organized themselves in the form of an informal association called *Gram Swarajya Jungle Sanstha* (Village Self-Ruled Forest Association) in 1979. Since then they have been protecting their forest. Meetings of the organization were held in which the villagers formed rules of forest use for all community members. Different patches of forests that were known by different names like *Ghodikhalche* jungle, *Mothya Devache Raan* and *Mukaji Devache Raan,* indicative of either the status (like a sacred grove) or its location (like 'forest below the horse like rock'), were made management zones for the community. For protection of the forest from outsiders, individuals from each household were sent as watchmen on rotation basis.

Mr. Baburao Dighore handed over the reins of the association to Mr. Janba Shivram Tondare, who took active interest in the working of the association. The indifferent attitude of the Forest department continued. Neither were the officials concerned about degradation of the resource, nor were they appreciative of the measures taken by Lakhapur. Total lack of coordination has been persisting ever since. Even after the department adopted 'participatory' approach in managing the forests, the relations have not improved. The community of Lakhapur continues to believe and claim that it owns the forest and its products as well as all other resources within the village. They feel that the forest has degenerated over the years due to neglect, inefficiency, and corruption amongst forest officials. Therefore, they have taken the task of managing and regenerating their own forest upon themselves. The community has similar feelings for other government officials as well. Recently it has taken over the primary school run by the *Zilla Parishad* (a government agency at sub-district level) by appointing its own teachers.

The 'Outside' Influence

The discordant note reached a higher pitch when Lakhapur community came in contact with a charismatic leader, Sant (Saint) Tukaram Dada who came to stay in a nearby place known as Adyal Tekdi, in early 1990s. The community was highly influenced by his ideas, and the saint managed to overshadow the local leaders. With a firm belief in decentralized decision-making, he prompted people to claim ownership of the resources of their villages, and make their own decisions regarding resource-use, and distribution of benefits. This philosophy rather appealed to a major section of the community since they already had bad experiences with the Forest Department. He appealed to the community in the name of judicial rights as well as social and religious responsibilities. Tukaram Dada continues to influence Lakhapur and a few other communities in the vicinity. His methods are religious, like daily prayers, moral classes for children in the community etc. on the 'Gandhian' lines. It is his teachings that led the Lakhapur community to consider itself the sole owner of the resource and thereby possessing the rights to use its product as it pleases. It became more defiant as well as adamant after it came into the '*gram swarajya*' (village autonomy) fold of Tukaram Dada in 1998. Lakhapur and Saigata had initiated forest protection work almost simultaneously, protecting their forests together. But the things changed with the entry of Tukaram Dada in the life of Lakhapur community. Saigata did not accept Dada's philosophy because it did not believe in confrontation, but in cooperation. The differences in beliefs of the two communities separated them and the cracks still show.

Structure and Working of the Forest Association

Since the inception of the forest association, its formal set-up involves an executive committee of nine members. Till now no women members have found place in this decision-making committee. Selection of the members of the executive committee normally takes place through elections, which are held every year in the annual meeting of the *gram sabha*, in which every member of the association participates. In case of an emergency, a meeting can be called but it has not been done since 1996. The term of the members of the executive

The forest association

Name of the Forest Association	Gram Swarajya Jungle Sansthan
Activities carried out by the members of the Forest Association	Plantation, protection, enforcement of rules
Formation of the association initiated by	Mr. Baburao Dighore
Year of formation	1970
Year in which legal status received	No legal status
Source of financial support for the association	Membership Fees

committee i.e. the President, Vice-President, Secretary, Treasurer etc. is fixed for one year, but the user group alone is authorized to remove these members. No external or higher authority is entitled to remove them. These members hold the positions in the committee on voluntary basis, i.e. they do not get remuneration either in the form of cash or kind. The elected members meet as and when the requirement of a meeting arises. So far, there has been no tussle for any position of the executive committee. Since its inception, Mr. Bhaskar Patil, Mr. Ramkrishna Raut, Mr. Sakharam Patil, Mr. Ramchandra Thaore, Mr. Shravanji Panse, and Mr. Sitaram Deshmukh (the current head of the association) have led the association.

All these members of the executive committee are also a part of the larger forest user group and have been ordinary members of the association at some time or the other. None of these executive committee members own any asset or capital that operates on raw material obtained from the forest. Most of the members reside in the village itself. The average education level of the members of the executive committee was found to be same as the average level of education of the other villagers, which was not very high. None of the members hold any position in other collective or government bodies except Mr. Sitaram Deshmukh who has been with the Malaria Eradication Department, Government of Maharashtra for a considerable period of time.

The association has been carrying out Forest protection activities continuously for the last two decades. The whole community of Lakhapur works voluntarily for the association. The members mainly interact with each other, regarding tasks to be undertaken within the forest such as harvesting, guarding, monitoring, maintenance etc. It was found that they could hardly take up any forest improvement or regeneration activities, over the years, due to paucity of funds. They have rarely planted seedlings or trees or cleared undergrowth. They have not built any fence to protect the forest, although watchmen have been employed for guarding the forest. During the past years, the members of the user group have not made any attempt to remove encroachments, or to create a nursery for seedlings, or to seek help from external authorities to improve vegetation growth, or to reduce harvesting levels of medicinal plants. This is mainly because the village does not have any financial assistance either from any private organization or the Forest Department. This is mainly because there are no private organizations functioning in this area providing this kind of help to the rural communities and Lakhapur community refuses to cooperate with the Forest Department. The only activity the village has undertaken over the years, for the forest has been clearance of dead wood and grass. In order to reduce their dependence on the forest and its products, the members of the user group have planted some trees, good for fodder and fuel, on individually owned agriculture land. Other than that they have not tried to develop any local technology to reduce their dependence on the forest. In the year 2000 the community had taken up some teak plantation on its own. The two main forest activities that however remain are regulating the use of forest products by community members and guarding the resource from poaching by outsiders. The association has coordinated activities related with maintenance, harvesting of forest products by determining time and quantity to be harvested, monitoring forest condition, and ensuring conformance to rules by forming rules and deciding penalty in case of infractions. The association now employs two paid guards throughout the year as against the voluntary protection work done by the community members. These guards are paid Rs. 600 per month out of the amount collected by all the members of the Lakhapur community.

Restrictive Rules

Although the Association carries out limited activities in the forest, it has set up several rules of use. The rules are framed by the majority vote in the *gram sabha*. No cognizance is taken of the rules laid down by the Forest Department. The harvesting rules differ from product to product. For fodder there are no rules regarding quantity or timing of harvesting. But the community often decides area of the forest from where it can be harvested. In case of timber, locals do not have any *de jure* rights, but Lakhapur community has certain regulations in this regard. The community members can harvest timber for house construction or for making agricultural implements, only after informing the *gram sabha* and getting its approval. In case of firewood, the local rules conform closely to the rules recognized by law, this being so, only by chance. The category of 'Protected Forest' under the Indian Forest Act provided for usufruct rights to the local community. These forests are usually demarcated areas closer to settlements for meeting the day-to-day needs of the nearby community, and are to be managed by the Forest Department. But as stated above, the Lakhapur community does not accept ownership rights of the Forest Department, and complete disrespect for the authority of the Forest Department.

The main forest products that are naturally available throughout the year, vary moderately in quantity from season to season.

Forest use and people's perception

Products harvested from the forest	Fuel wood, fodder, bamboo, timber for house construction, food supplements etc.
Total number of animals dependant for fodder	1510 (Cattle-350, Buffaloes-60, Goats-1000, Poultry-100)
Increase in the number of households in past 20 years	30
Villagers consider the forest	A sacred as well as an economic resource
Villagers rank their forest as	Somewhat sparse
Techniques adopted to reduce resource dependence	None

Fodder, though mostly available in the rainy season, is

difficult to harvest in this season. The forest use rules permit harvesting of fodder for six months of a year including the month of August when the availability is maximum. Although these rules exist, the community accepts that most of the time cattle are left for open grazing, putting a lot of pressure on the forest with 410 cattle and 1000 goats going to the forest everyday to graze. Since all the members do it, it is not treated as an infraction of rule. Timber on the other hand can be harvested every third year, and fuel wood, three quarters of the year, including the month of May when its availability is maximum. The dependence of the community on fuel wood is high. Nearly 80 per cent of its fuel wood needs are taken care of by the forest. One rule that is diligently followed by the community is that all the products are used strictly for self-consumption. Nobody is allowed to sell these forest products in the market.

In case of timber, total quantity to be harvested or used during a year is determined by making an estimate of the total availability of the product each year. The actual quantity harvested is distributed equally among all the user households, but in case of fuel wood, the total quantity is determined according to the requirement of each member household. These restrictions have been more or less the same for the past ten years for timber, but in case of firewood there have been some changes. The other restrictions are regarding locations within the forest from where timber or firewood can be harvested. All these decisions are taken in the *gram sabha* after consulting all the members. In case of both timber and firewood, a tree can only be cut down after it grows to certain minimum girth. This restriction has also not changed in the past 10 years. There are also restrictions on the means of transport of timber, which is that the product cannot be carried on bullock carts. It has to be physically carried by the members of the user group automatically restricting the amount thereby. No individual can go to forest alone for harvesting any product, he has to be in the company of at least four people. This has proved to be a very effective measure of monitoring. This rule structure apparently looks robust but does not deliver because of growing discontent among a faction of people.

Unreported Infractions

As reported by the community in Lakhapur, no infractions have been recorded so far in case of any of the products. However, this is a little hard to believe looking at the growing discontent and internal conflicts amongst the members of the association. Although no breaking of rules has been reported yet, penalties have been predetermined for any such possible infraction. The type of penalty for a user who breaks a harvesting rule for the first time has been decided, to be a fine more than one week's work. For the second time, the fine would be a permanent suspension of harvesting rights of the products, thus giving the offender no chance to repeat the offence. In case of any infraction, it has been decided that the appropriate penalties would be decided by the vote of the executive committee and the fine thus imposed will be collected by an official of the forest association to be treated as a source of revenue for the local association. The community is confident that if these penalties are imposed, the offenders will comply. If the offender does not pay the fine, the group has kept the provision of public rebuke to the offender and even social boycott. If the offender wishes that his harvesting rights be restored, he has to publicly apologize and give an assurance to the community members of non-repetition of the offence in the future. During the past 10 years, no infractions have taken place reportedly. All the members of the association are well aware of the rules and find them fair and legitimate.

Tall Claim

Not everything in the village seems to be as perfect as is claimed by the few respondents who spoke to us. During our meetings a small group of four or five villagers tried to give us the other side of the picture. Before they could tell us much, a heated debate ensued between the two groups. The dissatisfied group seems to be against the philosophy of isolating the village from government funded developmental works. They feel that their association too, before the advent of Tukaram Dada, was not only protecting the forest but also taking decisions at village level without antagonizing government officials. The dominant group on the other hand believes in total autonomy as far as the forest protection and improvement are concerned, and wants no interference, whatso-ever from the Forest Department, and is in constant disagreement with the other group which feels that assistance from the Forest Department will be useful for their forest. The dominant group, constituting mainly of people belonging to the Kunbi caste (the cultivators), constituting 50 per cent of the total population, is highly influenced by Tukaram Dada. The remaining 40 per cent Mahar (the former untouchables), and few Gond (numerically most dominant tribe of central India) and other caste groups are not in favour of the saint's teachings.

Local indicators of economic state

Local definition of wealth	Ownership of land, big house and possession of gold jewellery.
Local definition of poverty	Unemployment
Existence of economic disparity based on the local definition	Yes

This split in the community came out in open in 1998. The 73rd Amendment to the Constitution of India encourages decentralization and promotes 'self-rule' by the *gram panchayats*. It has provisions for transferring the rights of decision making, with regard to the resources of a village, to the *gram sabha*. Lakhapur was fast in establishing its authority on its forest by cutting down trees in the name of resource management and land-use policy. This action was apparently prompted by the external leadership. Legal action was initiated by the Forest Department, who arrested the culprits immediately. When Tukaram Dada could not help the arrested in any way, it fractured the community into two groups. The other group no longer trusted the external leadership. This led to internal conflicts affecting monitoring and rule-compliance. The association has become weak from inside because there are numerous unreported infractions. Rules for forest use have remained only on paper. The office bearers of the forest association, all belonging to the dominant group, claim that written records are being maintained regarding the history of the association, identity of the office-bearers, meetings and resolutions of the meetings, income and expenditure incurred by the association, amount of forest products harvested, distribution of forest products among users, list of individuals who failed to follow the rules of the association and were punished, conflicts of the association with individuals or other associations, and the type of punishment imposed. However, it is unavailable not only to outsiders like us, but even for the other community members.

The unjust methods of the office bearers are apparent from the 'equal contribution' principle adopted by them. The annual contribution by all the households towards the funds of the forest association is fixed at an amount of Rs. 100, irrespective of the economic status of the member. Though the amount to be paid is equal, equity is missing from the whole plan. The distribution of burden is not equitable between the households that own land and the ones who are marginal farmers or landless. Incidentally, all the office bearers of the association belong to the first category. The dissatisfied group being in minority and belonging to the economically backward class, can neither protest within the community nor can it approach the formal governmental agency like Forest Department or the Police. The traditions of not allowing others to interfere in the village matters has resulted in non-auditing of accounts, and of taking all the decisions by majority votes has rendered the minority uninfluential.

1. Symbol of Defiance: The board restricting Forest Department officials from entering the boundaries of Lakhapur

2. Encroachments done upon Lekha forestlands by the Mendha farmers

3. The unpretentious leader of the successful collective action in Saigata: Mr. Suryabhan Khobragade in the backyard of his modest house

4. Successfully and collectively ruined: The patch of Maral Mendha "forest-land" now chosen by the Forest Department for replantation

5. A Bamboo-made well-bunding: An example of indigenous technology in village Lekha

6. "*From conflict to collaboration*": Lekha & Mendha coming together at the initiative of Mr Kamlakar Jhanjhad (second from left)

7. The Gotul: The "Critical Event" leading to the community uprising in Mendha

8. Mr. Dewaji Tofa: The unanimously accepted community leader of Mendha

An Uncertain Future

It all began well. The villagers considered the forest as both a sacred and a good economic resource. The significant dependence of the villagers on the resource and shrinking of forest cover made them learn the value of the forest when it had already paid a high cost. This initially prompted them to protect the forest. The community accepted a self-imposed discipline of restrictive use, invested time and effort in the protection activity, and resisted commercial exploitation. But in the absence of support from the formal institution i.e. the Forest Department, it lacked in expertise and funds for development. In the absence of these resources, Lakhapur has not been able to qualitatively improve its forest.

Though the forest seems to be sufficient for the present needs of the community, its degeneration is likely in the near future if the present trend continues. Thus, Lakhapur is a clear example of what can happen in case of lack of coordination between formal and informal institutions. Though it claims to be a follower of Gandhian concept of *gram swaraj*, it is following it in a distorted manner. One cannot ignore the fact that the Gandhian idea of *gram swaraj* was developed in the pre-independence times. It is essential to reinterpret it in present context in the form of 'participatory', more meaningful cooperation/coordination between formal governmental set-up and informal efforts at local level. Defiance in case of Lakhapur has come as a reaction to inefficient management by the then forest officials. However, continued non-cooperation has only resulted in loss of income generating employment and opportunity to improve the forest quality. The Forest Department did show inclination to include Lakhapur under the JFM programme in 1992, but the villagers refused to co-operate with the Forest Department. Today, *Gram Swarajya Jungle Sanstha* has neither ensured sufficiency of forest products to the community nor has it succeeded in keeping the community as a cohesive group.

Basic details about the village

Elevation	256 metres ASL
Latitude	20°37′34.9″N
Longitude	79°46′48.5″EO
Nearest Town/Market/ Administrative Centre	10 km
Total Population	547
Sex composition	Males (53.58%), Females (46.62%)
Current number of households	110
Nuclear families	62
Literacy	33.82%
Dominant religions	1. Hindu (55%) 2. Bouddha (45%)
Dominant caste groups	1. Kunbi = 50% 2. Mahar = 40% 3. Gond = 2%

4

FORWARD LOOKING TRADITIONAL COMMUNITY OF MENDHA

"Collective action may take place through an organization, but also may form around informal social groups or even spontaneously to critical events."

—Dick, Raju & Gulati

In the year 1989, Mendha community decided to build a *Gotul* (a traditional community centre) in their village. Traditionally, *Gotuls* are built using Teak, the best timber, available in the adjoining forest. A few male members of the community collected the required timber from the forest, and the community started constructing the *Gotul*. Getting a wind of this, the Forest Department sent a guard to the site who confirmed that construction was indeed on. Two officials of the department were sent to talk to the people and dissuade them from the illegal act and surrender the valuable timber. They returned disappointed as the community refused to dismantle the structure. Then came the Range Forest Officer, who tried to convince the people and warned them of dire consequences if they didn't comply. The people did not budge even under the pressure of the Deputy Conservator of Forests, a senior officer of the department. At this point relations between the community and the department reached its nadir. The community members were ready for the worst. They had guessed rightly that the Forest Department would bring in the police. Strategically, the women of the community

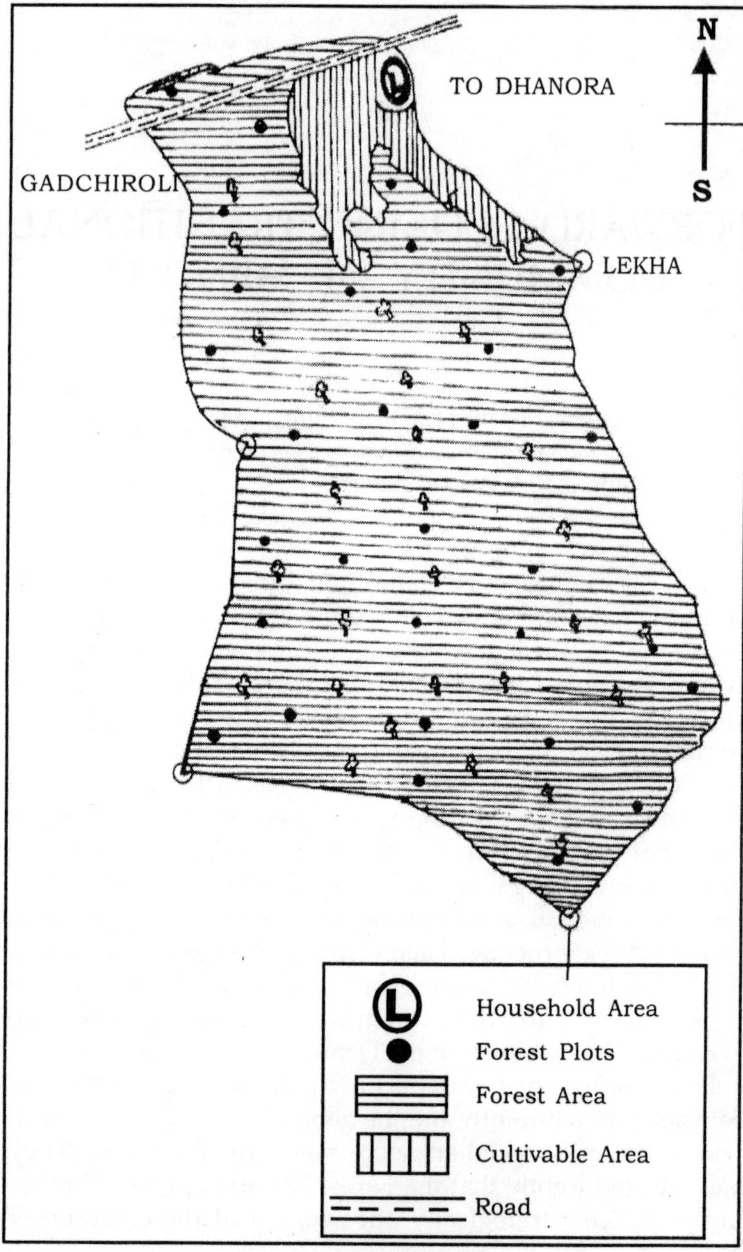

Map 4.1: Village Mendha

decided to face the police force, as it was believed that the policemen might hesitate to raise their baton on women. The police did come to the village to assist the Department with the task. The department staff, though successfully dismantled the structure and seized the timber, was unable to arrest anybody on the charge of felling the trees, as the whole village including the women and children came forward together for a mass arrest. Immediately after the government officials had left the village, people of Mendha, adamant on rebuilding their *Gotul*, went and extracted more timber from the forest and built the *Gotul* again within a day. The message was sent to the senior officers of the department that if they came again and destroyed the structure, it will be built again resulting is felling of more trees. The department wisely decided to turn a blind eye.

Certain tribal communities in India have a tradition of informally getting together every evening in a centrally located place, usually at *Gotul*. It is a place for elders to discuss day-to-day activities, for children to sit and learn about the community's history, their traditions and skills, and for young unmarried girls and boys to sing, dance and choose life partners. Ample evidence of the significance of *Gotul* is found in Verrier Elvin's study of Bada Madia tribe from central India. Over the years this institution is losing importance like so many other institutions of indigenous communities. At the same time, realizing the importance of indigenous knowledge, there are some communities who are trying to revive their lost heritage. The incident narrated above reflects the community's effort to revive this institution inspite of government resistance. The conflict is rooted in the competing claims of ownership of the resource. While all the forest land is owned by the government by law, the communities insist that they have been managing the resource traditionally for generations and the government ownership has not only deprived them of forest use but has also failed in protecting the resource from abuse. This strong conviction of today is the result of a chain of events that have taken place in last three decades.

Brewing Resistance

It all started when the community of Mendha, which is located in a forest-rich district of Gadchiroli, came in touch with a tribal leader who was consolidating his hold on tribal communities by demanding restoration of tribal rights on forests. Shri Lalsham Shah Maharaj, a descendant of royal Gond family started a movement for reviving tribal culture that was rooted in decentralized decision-making for resource management that ensured equitable sharing of benefits. The tribal tradition had received a severe setback when the forests were declared public property after the country's independence in 1947. In the changed circumstance, the contractors were given authority to fell trees according to the Forest Department's working plans, on which local communities had no control. Departmental supervision was often meagre resulting in felling of unmarked trees as well. The Mendha villagers, like many others, were mute witnesses to this degradation, believing that "you don't protect what you don't own". Their experience in community management had become redundant as they were brought under a state-regulated 'representative democracy', replacing their village authority. But they were also experiencing that the centralized forest management had failed both to protect the resources and to deliver the promised development. This was aggravating local communities into re-establishing their 'community ownership' by defying the national laws.

At that time another movement known as *Jangal Bachao, Manav Bachao Abhiyan*' (save forests, save human beings movement) during the second half of 1970s, led by a former tribal Commissioner of Madhya Pradesh Dr. B.D. Sharma, had spread to Gadchiroli. Under this movement the villages that believed in the philosophy of self-rule, were becoming pro-active. Some active non-government organizations (NGOs) and aware politicians, working hard to restore the long-forgotten traditions of community life, were encouraged by this movement that had swept the whole of tribal community in central India.

A Non-Enchanting Beginning

Mendha is a small, quiet village with 100 per cent tribal population. It was under the jurisdiction of one *malgujar* till 1955, the year when *malgujari* was abolished. Therefore, Mendha and its neighbouring villages, Kanhan Tola (another total tribal village) and Lekha were living as a unit, cultivating land in each others' village boundaries, and sharing forests as common land. Sometime in mid 1970s, one member of the Mendha community saw the workers of a forest contractor hacking off some branches of trees in Mendha's forest. He felt as if limbs of a human being were being cut. His request to the labourers to stop, fell on deaf ears. He came back to the village and shared his concern with others in the community. This small incident proved to be the spark required to set off the fire of resistance. A few days later Shri Lalsham Maharaj was to visit the neighbouring Kanhan Tola. Some members of the community were sent to go over and meet the leader. The *Maharaj*'s slogan for the movement was 'if the forests get cut, so will human beings be'. The *Maharaj* being from the area, used local *Gondi* language to communicate with the people. This was an added advantage. He explained to people how the government was being unjust to the indigenous people by troubling them over harvesting forest products specially timber for house construction. He encouraged the tribal youths to set up a 'Youth Forum' to take the message of 'protest' from village to village. The youths of Mendha took active part in this.

Farsighted Decisions

With the 'youth forum' giving the necessary impetus, the Mendha community started to get together sometime in early 1980s. One person from within the Mendha community who took the lead was Mr. Dewaji Tofa. He, like all others, was convinced of the fact that the tribal community as a whole would end up being losers by giving up some of the traditional practices. He decided to revive some of these consciously. The first and foremost decision was to assemble the community every day and discuss various matters in the most informal way. It is then that he asked the elderly in the village to talk about the past, the culture and traditions of *Gonds*, the state of forest, ownership patterns etc. During these assemblies the community decided to continue with some traditional practices that were sustainable and relevant to the current needs.

In many tribal communities it is customary to build houses out of grass and bamboo; teak (a valuable specie) is specifically not used. Similarly, grass is not cut for fodder until the seeds fall off, so that they can germinate in the next season. The water holes used by wild animals are not used for fishing. Using poison for fishing is prohibited. Instead, leaves of a particular tree are used, which make the fish unconscious and hence facilitate catching. This method doesn't harm the water quality or other insects. The knowledge behind these ancient rules followed by the tribals point to the fact that traditionally the tribals have used their wisdom for sustaining the forest and the diverse species dependant on it. In a sense, tribal culture being a "group culture", many day-to-day activities are undertaken in groups. Usually members of all households go to the forest together for collection of non-timber forest

Forest use and perception

Products harvested from the forest	Fuel wood, fodder, bamboo, timber for house construction, food supplements and wildlife.
Total number of animals dependant on fodder	447 (Cattle-225, Buffaloes- 41, Goats- 53, Poultry- 128)
Increase in the number of households in past 20 years	60
Villagers consider the forest	A sacred as well as an economic resource
Villagers rank their forest as	Very abundant
Techniques adopted to reduce resource dependence	Bio-gas plants installed, modern bee-keeping techniques learnt

products (NTFP). In earlier days before the banning of hunting, all the villagers hunted together and the catch was distributed equitably amongst all. Pregnant women got double their share, and members who could not join the effort, also got their due share. Even in those times when wildlife was available in abundance, hunting was restricted to customary celebrations only, and was not done as a routine.

Another very interesting practice reported from Mendha was that of *tulurmutha,* meaning a fistful of grain, mainly paddy. *Tulurmutha* is a practice where the village priests collect a fistful of grains from all the households after the harvest, at the village sacred grove. The grain is then distributed to all the households at the same place in a pre-sowing ceremony. This could have been an important practice resulting in preserving genetic biodiversity of seed pool. The Mendha community under the leadership of Mr. Tofa rejuvenated several of these abandoned practices.

Resisting the Temptation to be the Leader

During the movement led by Shri Lalsham Shah Maharaj and the *Jangal bachao, manav bachao* movement, one person who was associated with spreading the message of decentralized decision making, was Mr. Mohan Hirabai Hiralal. After a few years, failing health made the Maharaj take a back seat and Mr. Mohan was at the helm of affairs in spreading message by travelling from one village to the other, making tribals aware of their rights. He had been to Mendha on several occasions during this period of social unrest. He too had noticed the interest of Mendha community as a whole and Mr. Devaji Tofa in particular, in reviving the tribal tradition of collective action. In the whole process, he had a major influence on the villagers including Devaji Tofa. Mr. Tofa frequently consulted Mr. Mohan regarding various governmental policies, development programmes, and forest acts. Both the movements in this area gradually died, initial enthusiasm faded. But, Mr. Devaji Tofa as a local leader, kept the movement alive in the minds of his community. With the help and guidance of Mr. Mohan, he organized the Mendha community to strive for establishing their due rights on their village resources.

The discontent against the Forest Department was growing in the Mendha community. The tussle with Forest Department was regarding 'ownership of forests'. Whenever, contractors sent by the Forest department came to Mendha forest to cut timber, or to collect Tendu leaves, the people protested and drove them away. Gradually the community decided to rejuvenate its traditional *panchayat* for taking all the decisions pertaining to the collective interest of the community. After the major tussle with the Forest Department over the construction of the *Gotul*, which was the last straw on the camel's back, the Mendha people formed a Forest association called the *Van Sanrakshan Samiti* in 1989 to deal specifically with forest related issues. The most heartening thing, witnessed during the whole process is that it was a perfect opportunity for both Mr. Mohan and Mr. Tofa to establish their leadership in the village and take the reins of decision-making in their own hands. However, Mr. Mohan as a true catalyst restricted his role to an advisor, interfering only when asked for by the community. He wanted to make the community self-sufficient and not dependant on him for taking decisions. Mr. Tofa also did not intend to assume the role of a leader and corner all the benefits of a leader, including fame. He insisted on unanimity of decisions despite the process being highly time consuming. He convinced the community to retain the positive practices of the old tradition, and at the same time accept some modern values.

Present Association : Structure and Activities

Mendha village has a large forest area (1809 hectares) and a small population of 375. The villagers know the various forest patches as *Mahur Jar Palla, Wanar Chua, Elge Aher, Tahagad Metta, Saheban Metta* and *Jaganeke*. Each name in Gondi language has a contextual meaning.

Area

Total area	*1829.79 ha.*
Forest area	1806.49 ha. (98.73%)
Per capita forest	4.82 ha.

The community decided to form the *Van Sanrakshan Samiti* (Forest Protection Committee) basically to guard their forest from outsiders as well as to regulate internal use.

The Forest Protection Committee in Mendha represents the whole village, as two members from each household— a man and a woman, are members of the committee. The association at the time of its formation had no legal status, but now it is a registered body. Its present structure and process has remained almost same since its inception, except that it has signed a Memorandum of Understanding (MoU) with the Forest Department under the Joint Forest Management in 1996. The Executive body of the forest association of Mendha has 11 members out of which three are female. These women members have been a part of the general representative body for the last five years, but none of them have ever headed the association. The process of selection of the members of the Executive body of the association is by election, in which all the members of the association participate. The term is of five years for the elected members of the Executive body as well as for the President, Vice President, Secretary and Treasurer. Only the members of the general body have the power to remove the members of the executive committee, and no external or higher level

The forest association

Name of the Forest Association	Van Sanrakshan Samiti
Year of formation	1987
Activities carried out by the members for the forest association	Record keeping, enforcing the rules, protection etc.
Year in which legal status received	1996
Source of financial support for the association	Voluntary contributions

authority has a say in it. So far, the elections for executive body have been uneventful, with no competition for any position in the executive committee.

Since the inception of the association, the dynamic tribal leader, Mr. Devaji Tofa has been efficiently looking after the activities of the association. All the members of the Committee have been working on voluntary basis, and do not receive any payment or material compensation. All these elected members are also forest users, but none of them owns any capital asset, which require raw material from the forest for its operations. Most of the members of the executive committee live close to the forest i.e. within one kilometre of the forest. The education level of the members of the executive committee on an average was found to be very low, as it is in case of all other people from the community.

The first decision taken in this committee was to stop all neighbouring villagers from harvesting from their forest.

The tradition of collective, informal meetings where men gathered under a tree almost every evening, have been replaced by more formal meetings, taking place for the last two decades and decisions regarding forest use are being made after considerable deliberation. Monday is the day fixed for village meetings, when everybody gets together to discuss and solve the problems of the village and its forest. There is a provision for calling special meetings in case of an emergency and these have been called on several occasions. Whenever there is a conflict of interest, the matter is discussed in the meeting (formal meetings are called only if required, otherwise most of the problems are settled in informal meetings). The issue is discussed until a solution is found to the satisfaction of all. Usually the verdict given by the senior members of the *panchayat* is considered just and acceptable. Women are involved in every activity, and they share responsibilities in decision-making as well as in the implementation of those decisions. The implementation may also involve guarding forests at night, which the women do enthusiastically.

The Mendha Forest association has been active over the years and has undertaken many activities. There are frequent management and regeneration activities taken up by the members of the user group. Forest Department is also now aiding them financially and otherwise, under the JFM arrangement. Till 1996 the Forest Department did not play any role in the various activities of the forest in Mendha. After that, the department has coordinated with the *Van Sanrakshan Samiti* on some matters related with the forest. But there is no compromise by the community on matters like harvesting of timber by the department. Mendha community wants to regulate all the harvesting in its territory, and does not accept Department's interference in this regard. On the other hand, the department is objecting to the community's practice of harvesting NTFP for commercial purpose. Transfer of legal rights over some minor forest products from the department to the communities under the 73rd Amendment to the Constitution of India, is a recent phenomenon. The community has been harvesting the products for two decades now. Despite all this, the Forest Department is keen to be associated with the people of Mendha and undertakes various activities.

Coordination of Multiple Activities

The type of activities undertaken by the Forest Department and the Forest Protection committee are: operational activities, collective-choice activities, and constitutional-choice activities including harvesting/use activities. Operational activities mean the day-to-day actions of the forest users and others, as to when, where and how to withdraw products, or otherwise use the forest, how to improve the condition of the forest etc. They also include monitoring and imposition of sanctions and rewards, and exchange of information about the forest. Collective-choice activities involve the policy-making activities of the users, user group officials, and others about operational activities. Constitutional-choice activities are those that involve decision-making about who will make the rules. Harvesting or use activities involve day-to-day consumptive or non-consumptive harvesting, and use from or within the forest by the forest users. Here consumptive activities are those where the products are harvested either for self-consumption or for commercial use. Non-consumptive activities include worshipping, bird watching or taking a walk through the forest etc. In case of Mendha, the Forest Protection Committee carries out all the operational activities, like planting and other maintenance, forest protection and harvesting. The Forest Department and the Forest Protection Committees together enforce the rules. The policy regarding planting, protection of the forest, enforcement of rules and harvesting of the forest products, is all decided by the Forest Protection Committee. The Forest Protection Committee alone takes the decisions regarding harvesting or the usage of the products. The association takes active part in the management and protection of the forest despite the fact that the vast forest area of 1809 ha. supports only 80 households. The time and effort devoted by the Mendha community for protection appears exemplary, when we consider that these people have never faced any scarcity. The committee does not allow any outsider to harvest from its forest. Only recently, the neighbouring Lekha community was allowed to use approximately 200 ha. for meeting its fuel and fodder demands. It was done under duress. We will see this part of collective decision in the fifth chapter.

The community has high dependence on the forest not only for meeting subsistence needs but also for augmenting its income by trading certain forest products (forest produce being abundant), both at individual as well as at community level through cooperatives. The community fulfils 100 per cent of its fodder, fuel wood, timber and biomass or green manure needs and 10 per cent of its food needs from the forest.

Local indicators of economic state

Local definition of wealth	Ownership of land and house
Local definition of poverty	Being landless
Existence of economic disparity based on the local definition	Not much

The activities that this association has coordinated over the past few years are:

1. Harvesting forest products: For the community of Mendha, fuel wood is collected from trees like *Mushan, Kusar, Gundhi,* and *Ghonyat*. Bamboo, timber for house construction, wood for agricultural implements, medicinal plants, minor forest produces are the other important products harvested from the forest. For house construction purpose, trees like *Yen* and *Dhawda* are used;
2. Distribution of forest product among the members of the user group;
3. Sale of excess forest products;
4. Determination of timing of harvesting of forest products: The three important forest products i.e. fuel wood, bamboo and trees for timber, are available almost throughout the year, but the actual harvesting period differs. Fuel wood is actually harvested for only one quarter of the year; bamboo and timber are harvested only for one month every year. The quantity of the three forest products varies substantially from season to season;
5. Fixing quantity of forest product to be harvested by each household : In case of Mendha, the members of the user group harvest according to their basic requirement;
6. To determine the type of technology or implement to be used in harvesting these forest products: People use axes and sickles and no other tool to harvest. These tools are easily available in the local market. Use of these tools has helped limiting the quantity that is harvested;
7. Decide who is authorized to harvest forest products;
8. Determine type of use that can be made of forest product (including religious use);
9. Monitor forest conditions as well as conformance to rules;

10. Determine and execute sanctions (e.g. fines and punishments) for rule breakers;
11. Interact with higher authorities; and
12. Arbitrate dispute among local users.

In order to carry out these activities the work is distributed between the members of the executive body, six employees of the association and volunteers from each household. Record keeping, enforcement of rules, protection of the forest are some of the responsibilities of the members of the association. Each household takes turns in sending a member as a forest watchman, since it is not feasible to build a fence around the forest. Two persons guard the forest day and night and are paid from the committee's funds. The association has also taken up plantation of some valuable species with the help of the Forest Department to change the specie mix mainly to improve density and quality of the forest. Some of the members have built up a nursery under the department's assured 'buy back seedlings' scheme. Making use of many developmental schemes the community is simultaneously trying to reduce its dependence on the forest. For example, biogas plants have replaced many traditional stoves. There is a conscious attempt to increase household incomes by taking up forest related occupations without harming the resource. Some members of the Mendha user group took training in improved bee-keeping techniques and are now earning well from honey that is sold at the community centre. This has brought earnings to the association, as a result it is the only community in the area to own a telephone.

Ensuring Prudent Use

Although the average education level of the members of the association is very low, there is systematic maintenance of records. The records are related to:

1. The identity of the office-bearers ever since the formation of the association;

2. Minutes of the meetings, and recording resolutions made in each meeting;
3. Income and expenditure incurred by the association;
4. Details of contributions made by the members in kind;
5. Notes on the condition of the forest as perceived by the members;
6. Members who failed to follow the rules and those who were punished; and
7. Conflicts with other individuals or with the government, especially the forest department.

Other than these records, an attempt has been made to maintain the history of the association.

The major source of finances of the association is voluntary contribution by the members of the association i.e. the community as a whole, and the proceeds from the sale of NTFP. The main item of expenditure of the association is payment to guards. The association takes decisions regarding income and expenditure and no other organization has a say in deciding how the association spends its money. For the past 15 years or so, all the rules regarding forest use have been made in the committee meeting, which is attended by all the members. Every change in the rule structure is finalized only if it is unanimously accepted. Since the community members themselves formulate all the rules, there are fewer instances of rent seeking behaviour. A working committee, consisting of the senior, experienced members of the village, forms the traditional *panchayat* by unanimous decision. Till date there have never been elections for this purpose. This body is generally entrusted to ensure fair implementation of rules and regulations. This committee then constitutes a subcommittee i.e. the *Van Sanrakshan Samiti* (VSS) that regulates forest related issues.

The association has made rules regarding harvesting, processing, and selling of fuel wood, bamboo and timber, which affect the harvesting level or use of these products. These rules are mainly regarding the quantity of the product that can be harvested. The required quantity is determined according to the need of each household on the basis of the number of members in the household. Each household has to inform the *panchayat* and get its approval before cutting down a tree to build a house, make instruments, and so on. 'Head load' is a measure for fuel wood and 'cart load' for timber and bamboo. When converted in standard metric units, one head load of fuel wood or timber equals 20-25 kilograms, and one cart load of bamboo is equal to 200 kilograms. There are also rules regarding timing and purpose of harvesting. Most of the forest products are to be used only for self-consumption. These rules have not changed much over the past 10 years. There are also restrictions on the location within the forest from where these products can be harvested. Grazing is strictly restricted to demarcated land and rule violators are punished. This restriction has changed over the past 10 years according to the availability. Restrictions include the form of a product that can be harvested, for example, in case of fuel wood, the harvesting is limited to dead wood only, and in case of fodder only mature grass can be collected. There are rules that include the type of transport to be used for removing the harvested product from the forest. Either the members carry the harvested product physically or use bullock carts. No change is found in these restrictions over the past 10 years. Forest produce is allowed to be harvested for sale, only after fulfilling the requirement for household consumption. Mainly, non-timber minor forest products are collected and sold, but the decision to this effect is not taken individually but by the *panchayat*. *Gram Sabha* (general body) has the right to decide whether a forest product is to be sold individually or collectively through its cooperative. The sale of NTFP has been one of the largest sources of income for the committee, yet there has never been over-harvesting of the products.

Thorough Acceptability

The rules and regulations of VSS are clear and easy to understand. The user group finds them fair and legitimate.

These rules have been made flexible enough to deal with emergencies. It is interesting to note that even after entering into the JFM arrangement with the Forest Department, the rules followed are the ones formed by VSS because these are considered fair, efficient, sustainable, and democratic, unlike the provisions in JFM. It is so for more than one reasons, e.g. the use of forest produce for household consumption has been given top priority; the regulated access to the resource has provided assurance to all economic groups; strict vigilance has successfully kept illicit felling by outsiders to a bare minimum; the forest-based household needs of fuel wood, fodder, small timber, and other NTFP are taken care of within the village boundaries thus saving money, effort, and time; and the community has developed a conflict-resolving mechanism, through its daily informal get-togethers and commonly shared norms, which is acceptable to all and is considered the least expensive option.

The individuals reportedly follow these rules and restrictions, and there have been no infractions resulting in 100 per cent compliance of rules. Although, no instances of infraction(s) have been reported regarding forest product harvesting and related issues, tough penalties have been decided in case of need. In keeping with the traditional systems, an offender is reprimanded verbally for the first violation, and only on repeating the violation that she/he has to put in extra labour as a payback to society. These social sanctions have proved effective, and the community reportedly has never had to force monetary penalties. The people of Mendha are very confident that all the decisions taken by the *gram sabha* are complied with. As the *gram sabha* is a powerful structure in Mendha, it is associated with decision making not only regarding forest, but also all other aspects of public interest. And it has succeeded in imposing fines and penalties, monetary or otherwise. Because of this, the community is confident of success in case of forest association as well. During the past two years, government officials have been called only to attend the *gram sabha* meeting, as the penalties are decided and imposed by the community.

The association has a written statement of mission and objectives prepared by the members of the community. These are not based on any set of rules provided by the government. The users alone have the power to change the statement, only if it is done unanimously.

Crux of Success

In addition to traditional practices, which ensured equity and sustainability, Mendha community has opted to remain flexible and has adopted some new norms like encouraging women to participate, experimenting with improved methods of cultivation, venturing into modern, but environmentally sound avenues of self-employment like bee-keeping etc. Presently, all the major decisions regarding forest use are taken by this committee, and are applicable to forest within the village boundary. Conflicts in the group have more or less died down and everybody seems to work in a united way. The villagers consider the forest not only a sacred resource but also an important economic resource. Discussions with the local villagers strikingly brought out their sense of ownership of the resource despite the fact that legal ownership of the forest rests with the Forest Department. It is because of this that the villagers have taken on a considerable level of responsibility. The villagers have regularly deliberated on its rule-forming mechanism, and to date social fencing has proved to be extremely successful. Now the forest belongs to the community, and is protected and maintained by them.

It is also important to note that the unity of the village is not restricted to issues related to forests alone. All the actions of individuals that affect the collective life are under the purview of the *gram sabha* and traditional village *panchayat*. This has made the process of 'association' itself very strong. Insistence of 'unanimity' of all decisions has in a way ensured compliance, though at an enhanced cost (Ghate, 2000). Acceptance of, and regard for modern norms, and, most importantly, maintaining the component of flexibility in the local/informal institution are some of the important factors in the success of this traditional leadership. Democratic values seem to be inherent in this community. There is a fine blending of traditional values and modernity, which is an indicator of political modernization. Involvement of women and all other members of the community in decision-making, a significant aspect of democratic decentralization, were accepted in Mendha much before an amendment to this effect was made in the Indian Constitution.

The ground situation is such that aspirations of the locals and the formal governmental institutions like Forest Department, are not compatible. Due to judicial superiority of the State, most of the villages have given up their traditional rights over the resource by accepting the rule of union government. But Mendha community has decided to stand up against it. There is a measure of defiance on their part as they find the bureaucratic attitude of the Forest department rigid. In general, the awareness level of the community of Mendha is high and their willingness to find out more has encouraged them to take up several studies with the help of experts from outside. For the present research as well, members of the association wanted to know the purpose of our study. It was only after they found our answers to their satisfaction, that we were given full cooperation. For visitors like us, although it is easy to conclude that this is due to the able and catalytic leadership of Mr. Mohan Hirabai Hiralal, some credit for the success of collective action should also be ascribed to the homogeneity of the group and its tradition of group decision-making. Historically, *Gond* dynasty has ruled and provided leadership in this part of region for a long time. Yet, not many *Gond* villages in the vicinity have come up with collective action or have provided leadership. It is surprising to see that in fact there is no one leader in the village. There are many who would talk to the visitors with equal confidence and efficiency. Mendha often has visitors coming from all over India due to lot of publicity it has received in recent years and it is not always that Mr. Tofa is available for providing information. But one can talk to any member of the community and extract equally reliable information. And here is the crux of its success. The educated and motivated leadership that came in the form of Mr. Mohan did not stunt the growth of the community. He did not give ready solutions for any problem, instead encouraged and enabled the community to find its own way out. Even Mr. Tofa, who too could easily have remained the leader of the association for ever, had he not allowed and insisted on developing the second rung of leadership. He chose to transfer the leadership position from time to time. Mendha in this sense is a unique community that has learnt to blend the modern with the traditional, resulting in a 'win-win' situation. No wonders that the 'Mendha model' is attracting and inspiring many neighbouring communities.

Basic details about the village

Elevation	247 metres ASL
Latitude	20°14'41.4"N
Longitude	80°17'04.8"EO
Nearest Town / Market / Administrative Centre	3 km
Total Population	375
Sex composition	Males (57.07%), Females (42.93%)
Current number of households	80
Nuclear families	80
Literacy	26.6%
Dominant religions	Hindu (100%)
Dominant caste groups	Gond (100%)

REFERENCES

Elwin, Verrier (1958). *Leaves from the Jungle: Life in a Gond Village*. New Delhi: Oxford University Press

Ghate, R. (2000). "The Role of Autonomy in Self-Organizing Process: A Case Study of Local Forest Management in India", Working Paper. Bloomington: Indiana University, Workshop in Political Theory and Policy Analysis.

5

GOVERNING THE COMMONS: A COMPLICATED TASK IN LEKHA AND MARAL MENDHA

Many cases of the non-emergence of self-governance can be due to difficulties at the level of problem recognition and placement on an agenda.

—Bonnie McCay

Lack of initiative in self-governance can be a result of various factors. At times some people may simply be unaware of environmental or common-pool resource problems. Others may be aware but not convinced that they can do anything about it or if the effort is worth it in the given situation. In some other situations the problem may lie in the inability of the community to come up with solutions that are acceptable and reasonable to all the members of the community. Lack of required funds could be a problem in a few cases. Then there could be other competing obligations or fear of reprisal from those with other interests, and inability of the community to resolve this conflict. In this chapter we present two such communities who have not been able to put the act together. Lekha community, neighbouring the Mendha community, has been trying for past two decades to solve its commons problem through mutual understanding and collective decision-making. It is only recently that it has achieved some success. Maral Mendha community, located close to two communities namely Saigata and Lakhapur who have been

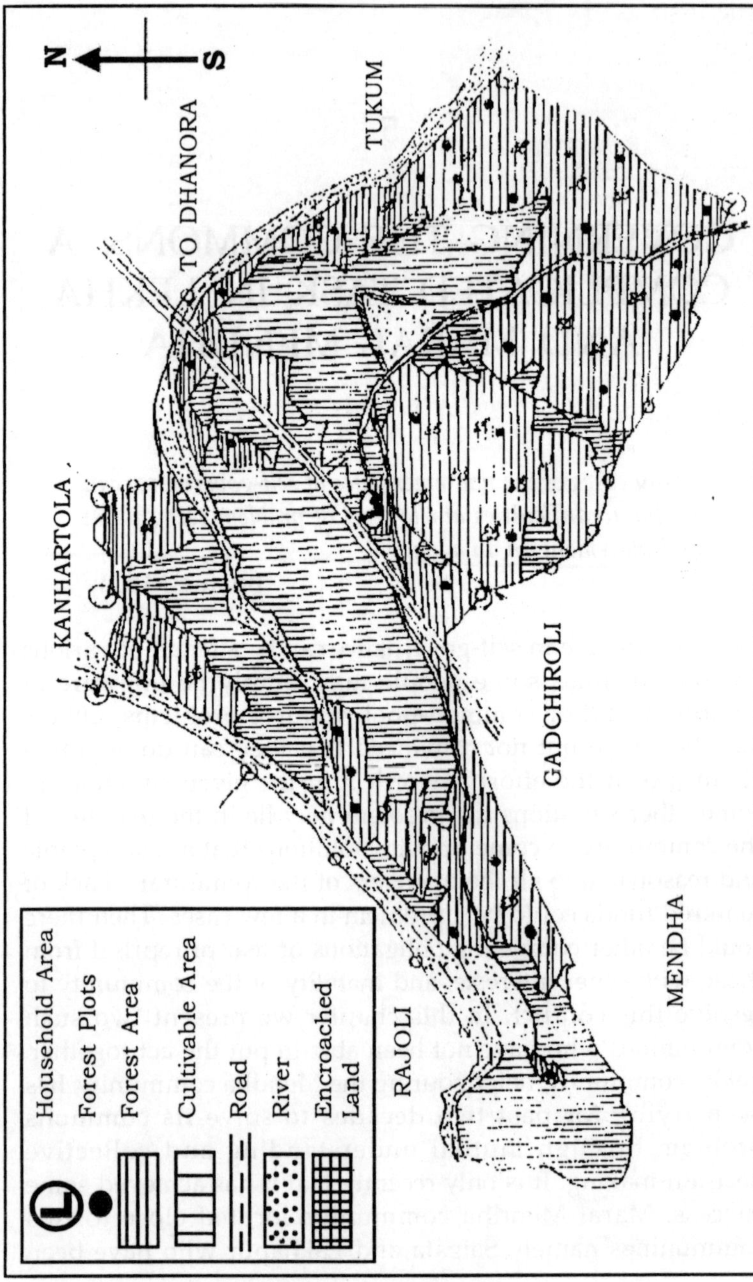

Map 5.1: Village Lekha

successful in protecting their forests to a large extent, is an extreme case of lack of collective action. It is difficult to believe that in fact Maral Mendha community was the first to get together, even before Lakhapur and Saigata, and had even formed an association for protecting the forest within its vicinity. But the effort was soon given up. There are lessons to learn from the experiences of both these communities.

Tricky Situation for Lekha

Lekha village seems to have passed through many phases before it could achieve some understanding, shared by the majority of the community members. The present structure of the organization is a result of the efforts of a small group of like-minded, school educated, youths of the village. Their efforts involved frequent changes in their strategy to make Lekha community realise the need for collective effort to save their forests. Abundant forest and low population pressure in whole of tribal region in central India has kept the local population away from the worries of degradation for a long time. It was true in case of Lekha as well. However, after its neighbouring Mendha community started restricting access to their forest, Lekha started facing resource scarcity.

Although it has been mentioned in the first chapter, we would like to restate the historical background of the village boundary demarcation of Lekha. Till 1955 the *malgujari* system, established by the British, was in operation. Under this system, the government used to appoint a rich landlord for collecting agricultural tax from cultivators, which gave him the title of *Malgujar*. He was also responsible for deciding land-use under his authority. The *malgujar* being a government official, often enjoyed the status of a *maharaja*. Village Lekha came under a *malgujar* who also had Mendha and Kanhantola under his jurisdiction. In case of this cluster of three villages, the topography is such that a major chunk of the fertile land is around a small stream that passes through Lekha. Therefore, most of the land that lies in the jurisdiction of Lekha is not only owned and cultivated by the farmers of

Lekha but also that of Mendha and Kanhantola. On the other hand, a vast tract of forest land in the south of Mendha has rocky terrain and, is not very good for cultivation. Hence, it was earmarked for forest related needs of all the three villages. This understanding continued even after the abolition of *malgujari* system, after which the three villages were demarcated as three separate units, and it did not pose any problem till as late as 1989. That was the year when the Forest Protection Committee of Mendha was formed. It is then that the Mendha community became conscious of its village boundary and denied access to its forests to the other two villages.

	Area
Total area	579.47 ha.
Forest area	279.51 ha. (48.24%)
Per capita forest	0.47 ha.

Lekha community was at a disadvantage because while the cultivated land had clear tenure rights in individual's names, the Lekha people had no such right on the forest lands of Mendha. The extent of forest dependence was immense in case of both the villages. The villagers extracted fuel wood, timber, wildlife and other non-timber forest produce from the forest. However, in the changed circumstances, farmers from Mendha continued to till their lands falling in Lekha's territory as a right, but members of Lekha community could not harvest from the forest.

Basic Information

Lekha village is in Gadchiroli district of Maharashtra state and is situated on the Gadchiroli-Dhanora highway. It is at an elevation of 248 metres above the mean sea level with latitude of 20° 14′ 57. 4″ North and a longitude of 80° 17′ 15.5″ EO. It has a common border with the Mendha village from the south side. The local languages spoken are Marathi and Gondi. The village has a population of 595 individuals. Out of these only 28 are nuclear families. The community has grown gradually. There were only 80 households 20 years ago, now there are 114 with 288 female (33 children) and 307 male (39 children) members. Although Hinduism is the only religion followed by the members of Lekha community, it is divided in many sub-groups on the basis of castes. The main castes are Mali, Gond and Dhivar, constituting 39 per cent, 26 per cent, and 11 per cent of the population respectively. Only 104 individuals (17 per cent) in the village are literate, though compared to the literacy levels of surrounding villages, it is quite high. Amongst these, 60 individuals have passed primary school, 11 have passed high school, and 2 are college graduates.

Agriculture is the dominant occupation. The main crops grown here are paddy, and pulses like Moong and Urad. Some farmers who have their lands closer to the stream also grow Chilly, as a cash crop. 52 households own agriculture land, while 5 households are landless and till others' land. Two households are engaged in commercial operations like running a grocery shop etc. Two women also own land, which is rather uncommon. Out of the 52 households that own land, 35 households do not possess land enough to meet their subsistence needs. Some of the households consume their own food crop that lasts throughout the year. A few households are able to sell their foodgrain in the market either because they have surplus, or because they are compelled to sell it for want of cash. The prevailing wage rate is as low as Rs. 25 for women, and Rs. 35 for men. But the villagers have no complaints because they have worked at daily wages as low as Rs. Three for women and Rs. five for men, five years back.

Local indicators of economic state

Local definition of wealth?	More paddy grown, more livestock owned
Local definition of poverty?	Not having agricultural land and livestock
Based on the local definition did a disparity exist in the economic status of the people?	No

Most of the houses in the village have houses mainly made of concrete with tiled roofs, indicative of good economic condition. But there are also few thatched houses with walls made up of bamboo and mud. Being in such proximity of the forest, all the households use wood for cooking. The village is just three kms from Dhanora, which has a market as well as governmental administrative centre. It takes them only 30 minutes to get there in the dry season and 60 minutes in the rainy season, by bicycles.

During our first visit to Lekha in 1995, we had found that the forest association was hardly there. It was active intermittently, lacked coordinated work, and was in serious need of efficient leadership. Patches of forest were being continuously cleared for cultivation along the riverside. The villagers themselves had categorized their forest as "somewhat sparse". But a group amongst them had not given up effort of bringing the community into a common understanding. The *Van Sanrakshan Samiti* of Mendha intensified its vigilance in the forest and altogether prohibited the entry of anybody other than the Mendha user group. This had led to a vicious rivalry between the two villages. It grew to such an extent that the residents of one village did not enter the boundary of the other. As mentioned earlier, this was when Lekha started facing difficulties in procuring fuel wood and other forest products of daily requirement. Since the village had deteriorated its own forest to an extent where they could not fulfil all their needs from the forest that was legally theirs, they understood the importance of a well-kept forest. Therefore, in the early 1990s, the people of Lekha decided to act towards saving their own forest from destruction. That probably is the reason why attempts at conservation started here long after it had already started taking shape in many other villages.

Being non-tribal, the Lekha people do not see their forest as something sacred that they would like to worship, but consider it an economic resource that provides them with their daily requirements in the form of fuel wood, fruits, fodder for the animals, herbal medicines etc. When they formed their association the chief aim was supposed to be protection of the forest to assure supply of daily requirements like fuel and fodder. While forming the committee, the plan was that the whole group would participate in the process of rule making through discussions, suggestions, arguments etc. But the internal conflicts happened to be too many and the leadership too weak to be able to make the association useful.

The next visit to Lekha was a pleasant surprise for the team, as previously the villagers were seen to have lost hope of success in their efforts. Now the village had found an emerging leader in shape of a young Mr. Kamlakar Jhanjhad, who adopted unique techniques to end conflicts within the group and to resolve the rivalry between the two villages, Lekha and Mendha.

Forest use and perception

Products harvested from the forest	Fuelwood, fodder, bamboo, timber for house construction, food supplements etc.
Total number of animals dependent on fodder from forest	510 (Cattle-200, Buffaloes-10, Goats- 200, Poultry- 100)
Increase in the number of households in past 20 years	34
Villagers consider the forest	An economic resource
Villagers rank their forest as	Somewhat sparse
Techniques adopted to reduce resource dependence	None

A Long Battle with Oneself

Lekha community was oblivious of the problems that might arise out of indiscriminate harvesting in the forest till a decade ago. When they did understand the problem, they were unable to work out effective solutions. The community, being heterogeneous because of people coming from different places and settling down there at different periods of time, did not have a traditional *panchayat*. The only institutional structure was that of local government's lowest unit, also known as *panchayat*. Lekha being a small village, it is clubbed with Mendha, and the two form what is known as *gat* (group) *gram panchayat*. Till late, this body had a very limited role to play, with no financial authority in deciding the issues of the community. The head of the *panchayat* was to be elected alternately from Lekha and Mendha by the two communities combined. Usually the *sarpanch* (head) of a village, who is elected as the head of the *Gat Gram Panchayat*, is an elderly, respected person from the community, irrespective of his education, knowledge, and interest in solving the problems of the community.

This kind of leadership failed to give proper direction to the aspirations of some who advocated collective action to save their forests. The group of interested young men was unable to convince the then existing leadership that the community needed to take initiative in solving its own commons' problems. The elderly *sarpanch* insisted that the community could not help the situation as the forest belonged to the government and it was government's responsibility to protect the resource.

The hardships caused by deteriorating state of forests, massive encroachments in the forestland belonging to Lekha, were enhanced by denied access to Mendha forest. These concerns were shared with one and all in the community. However, the established leadership was not ready to take up the issue and the persons who were ready, were not accepted as leaders. Mr. Kamlakar Jhanjhad did make several attempts to bring the community together by calling meetings, discussing various aspects of the forest-related scarcity experienced by the community etc., but was unable to convince them into a collective action. He was in constant touch with the developments in neighbouring Mendha, watching them making decisions, negotiating with forest officials, building a robust institutional structure through rule-making, sanctioning, monitoring etc. He also shared this information with his own fellow villagers. But even this failed to give sufficient impetus to the movement of collective action. In 1995, the forest department had approached the *Sarpanch* of Lekha, Mr. Mahagu Uike with an offer to bring the village under Joint Forest Management, to which he showed no interest. The villagers were apparently under the misconception that JFM was not meant for non-tribals. The concerned Forest department officials, who ought to have educated them on the terms and provisions of the scheme, were not very keen to put in the necessary time and effort.

After successfully convincing a few young people in the village, Mr. Jhanjhad suddenly realized that his growing popularity amongst the villagers was being perceived as a threat by the formal leadership which was making them all the more adamant. He realized early that it was important for the survival of the movement that the internal rivalry be resolved. This made him change his strategy. He started giving credit of his actions to the *sarpanch*. He would make the visitors, coming to visit him for some work, go to the *sarpanch*. He started telling the visitors of all the good work, including forest protection work, done at the initiative of the *sarpanch*. Though he himself was leading the movement, he let the senior members take the credit just in order to keep the movement alive. Instead of trying to fight it out with the opposing members, he started taking their opinions whenever it came to making decisions regarding other developmental schemes etc. He suggested their names for the higher administrative posts of the village. He had voluntarily kept himself away from all the powerful posts to keep the egos of other community members satisfied. He did not even try to become the president of the *Van Sanrakshan Samiti* of the village. This served his purpose well and the *sarpanch* gave him the power of attorney to deal with forest related issues. A few years later, this proxy leader was able to come out in open and directly confront the problems of managing the forest commons, after getting elected as *oop* (vice)-*sarpanch* of the *Gat Gram Panchayat*.

Collective action after a Long Wait

In 1997, when Mr. Kamlakar Jhanjhad became the *oop-sarpanch*, Lekha approached the forest department to get associated with Joint Forest Management, surprisingly the Department has not yet responded. He then openly tried to unite the people for the cause of protecting the forest. He also wanted to put an end to the rivalry with Mendha. Today, at the end of the five years of Jhanjhad's tenure, the situation has visibly improved. Conflicts within the Lekha community have almost vanished, and the work of forest protection has begun.

Negotiating Effectively

Mr. Jhanjhad's foresight and his capability to negotiate have brought Lekha a new hope of dealing with its forest scarcity issue. His leadership quality is once again proved by the success in solving the twin problem, one of hostility between the villages, and the other of reviving the dying forest protection movement.

As Lekha had already degraded its forest to quite an extent, and whatever remained, wasn't sufficient for the population's needs, it was first important to find some source of fuel wood and fodder for the community before the conservation movement could proceed. As mentioned earlier, according to the new boundaries, Mendha had received more forest than it required, while Lekha had got just 200 hectares of it. Lekha had allowed the landowners from Mendha to continue with their agricultural activities not only on the land owned by them but also on the encroachment lands in Lekha's forest. Mr. Jhanjhad was the first to think about negotiating with Mendha on these lines. Since the encroachments on the forest were illegal, Lekha did not find anything wrong with a bit of arm-twisting. It asked the Mendha community to either give them access to its forest land, or vacate the encroachments in Lekha's forest. Mendha had to yield to the pressure, as land is important for the livelihood of majority of the Mendha residents. It took both the parties some time to arrive at the terms of negotiation. It was finally decided, that the Mendha people would continue to cultivate their lands in Lekha and in return would give 200 hectares of forest to Lekha for exclusive use. It is now the responsibility of the Lekha community to guard this forest. Taking this extra bit of forest into consideration, Lekha now has a little more than 400 hectares of forest, which would suffice for the daily requirements of the village if managed well. In other words it can be said that a positive trade off has taken place as, while Lekha has been shifted to a higher indifference curve through this bargain, Mendha has not become worse off. Thus having the basic needs fulfilled, Lekha can and is thinking of conservation with a fresh perspective.

Lekha, under the accepted leadership of Kamlakar Jhanjhad, has applied for the registration of its *Van Sanrakshan Samiti* to the forest department once again in 2002. This reveals its desire to be a part of JFM as a formal management institution.

The forest association

Name of the Forest Association	Van Sanrakshan Samiti
Activities carried out by the members of the Forest Association	Protection
Formation of the association initiated by	Mr. Kamlakar Jhanjhad
Year in which legal status received	Not yet received

Choosing the Right Agent of Change

The important causes of the initial failure, the continued efforts, and the recent hints of success are more or less similar. While, weak leadership was responsible for the initial failure, the continuation of the efforts and the recent results can be attributed to efficiency and farsightedness of leadership. Leadership that was forced by the government in the form of a *sarpanch* lacked vision while the person with vision was not acceptable to the community, which on its part was not ready to experiment. If only, Lekha had a capable and established leadership during its first attempt at forest protection, the village could have been registered under JFM by now and could have availed the facilities provided by the Forest Department under the scheme. As help was not sought from the Forest Department at that time, no management or plantation activities could be carried out due to insufficient funds. The local leadership that the Forest department could have provided to this village was also absent. The favourable developments of the past couple of years, such as the agreement between the two villages, removal of internal conflicts within the group and the willingness of the village to come under JFM can 'therefore' be attributed to able leadership. Though the attitude of the forest department seems to be unchanged, emergence of internal leadership has made the alternate self-governance, a possibility. The unique thing about Lekha village is that a few residents of the village have been tirelessly trying to organize the community for protecting its forests for more than a decade now. They have encountered many constraints, have failed many times but are still not ready to give up. The village, under the strong leadership of Mr. Jhanjhad, is striving to make collective action possible despite the heterogeneous nature, and limited, deteriorated forest under the community's jurisdiction.

Basic details about Lekha village

Elevation	248 metres ASL
Latitude	20° 14'57.4" N
Longitude	80°17'15.5"EO
Nearest Town/Market/ Administrative Centre	3 km
Total Population	595
Sex composition	Males (51.60%), Females (48.40%)
Current number of households	114
Nuclear families	28
Literacy	17.49%
Dominant religions	Hindu (100%)
Dominant caste groups	1. Mali (39%) 2. Gond (26%) 3. Dhivar (11%)

A Perfect Example of 'TRAGEDY OF COMMONS'

"In an extreme case, open access occurs when there are no controls on the access to and the use of the resource. This leads to competitive and unrestricted entry and exploitation of the resource by all parties, which usually result in too rapid exploitation and resource rent dissipation" (Libecap, 1995)

Maral Mendha seems to be a typical example of this phenomenon. It is a village that has 71.46 hectares of land as forest land in its village boundary that was once wooded with many precious trees, and in just three decades it is all virtually gone. Except for a two hectares patch, all the forest land has been encroached upon, and at present, is under cultivation. It is difficult to believe that the open land used for grazing was initially a vast and rich forest. It doubtlessly, is an extreme case, where the community has, within a brief period of two decades done away with its common property completely by competitive exploitation and encroachment. It seems as if the community has succeeded in collective action for 'ruining the commons'.

Area

Total area	281.49 ha.
Forest area	78.12 ha. (27.75%) (Officially)

As is true of most of the villages in India, very little is known about the history of this village since no written or published records could be found. From the accounts of the villagers it seems that the earlier generations of the natives of the village had wandered into the village sometime in the early eighteenth century. They were in search of fertile agricultural land and had settled down close to a water tank. The settlement got its name Maral Mendha because it was dominated by the people from *Marar* (*mali* – a caste of vegetable growers) community. Along with the subsistence crops, the community, at that time, was able to grow onions and sugarcane; two crops that need a lot of water to grow. This became possible as Maral Mendha had sufficient irrigation

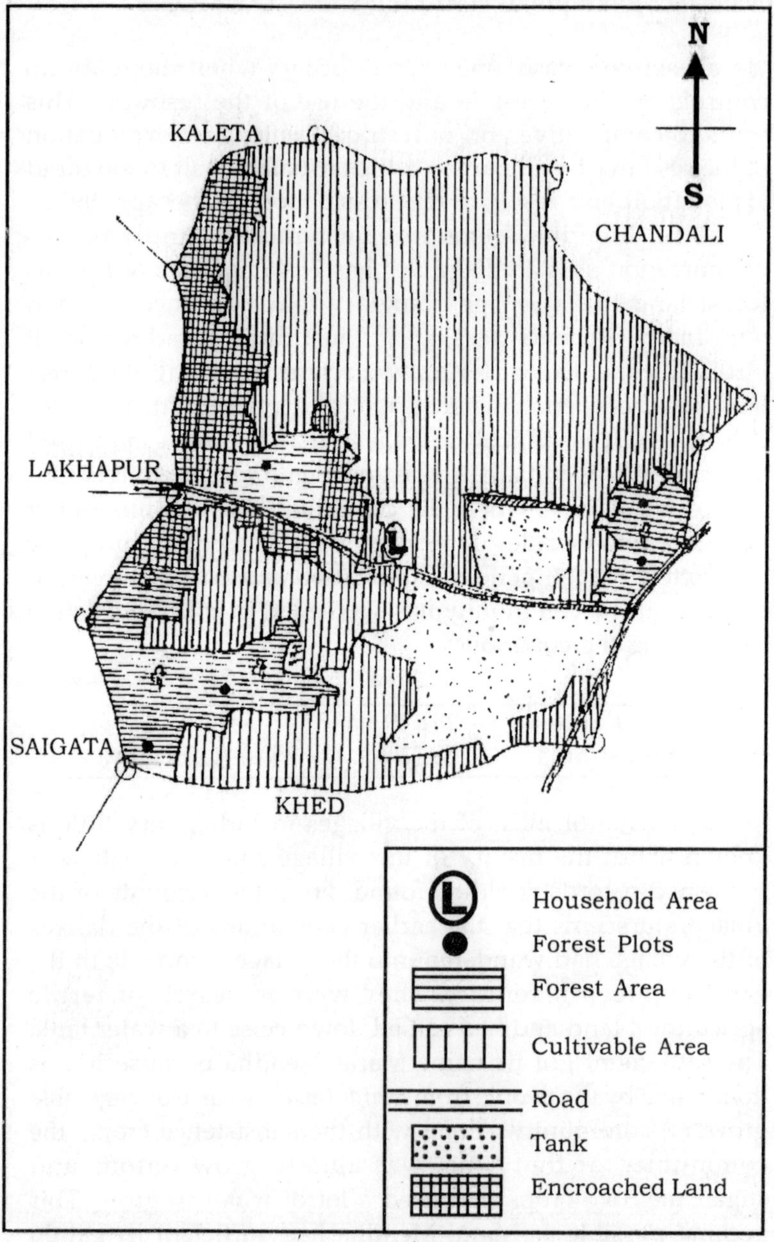

Map 5.2 : Village Maral Mendha

water available in its huge 600-hectare tank. Since Vegetable crops brought them immediate cash income, the Maral Mendha community started encroaching upon the forest land, converting it into agricultural land, thus discounting the future heavily. Those who owned land, cleared the adjoining forest to increase their farm size, and those who did not, simply cleared a piece of forest land and started cultivating it. Their vision was myopic regarding the long-term benefits arising out of existence of the forest. They were more interested in agriculture and its immediate gains as compared to the distant and unsure benefits associated with the forest. This self-interest encouraged them to systematically clear bits of the forest of the village.

Forest use and the community's perception

Products harvested from the forest	A little fodder. Forest non-existent
Total number of animals dependant for fodder	872 (Cattle-200, Buffaloes-40, Goats-130, Poultry-500)
Increase in the number of households in past 20 years	85
Villagers consider the forest	A hindrance to the growth of agriculture
Villagers rank their forest as	Very sparse
Techniques adopted to reduce resource dependence	None

Even after continuing this practice for many years, they were still left with some forest that provided them with their daily necessities of fuel wood and fodder for the animals, in addition to the forest area of Saigata and Lakhapur.

According to the forest map of 1977-78, out of the 71.46 hectares of total forest area under the village territory, 70.83 hectares was protected forest, and 00.63 hectares was shrub forest. The villagers remember the forest, which at that time had many valuable timber and fruit species. Although no physical evidence is found anymore, the elderly residents of Maral Mendha remember having valuable species like teak in its forest. They even boast that although the total forest area was lesser than that of Lakhapur and Saigata, the quality of forest was much superior. They feel that it is this feature of the forest that attracted more timber smugglers to Maral Mendha, contributing to its degradation. Irrespective of this, no measures were taken to protect the forest from the encroachments by the user group nor was any attention paid towards saving it from external harvesters. It is true that the two villages namely Uchali and Kalita, that share common boundaries with Maral Mendha, had little forest area allotted to meet their forest-based needs and, therefore, it is highly likely that Maral Mendha had to face the pressure of this excess requirement. The present reality, however, is that most of the land demarcated as forest in government record is encroached for cultivation, and not a single tree is found on the remaining two hectares.

An Attempt of Getting Together

The systematic cutting of the forest continued till mid 1970s. While more and more area of forest was being brought under agricultural activity, the community had started facing greater difficulties in availing fuel wood and fodder. It is around this time that the neighbouring communities of Saigata and Lakhapur villages had started protecting their forest, thereby denying access to Maral Mendha. A small patch was still under forest cover in Maral Mendha at that time. This led a few villagers to raise the question of forming a forest association in 1979. As a result, an informal forest association was set up to make forest related decisions and save whatever was left of the forest. But the effort failed miserably, and within two years of its inception, the association became defunct. In this community, majority was used to taking two crops a year, and saw huge potential in converting forest land into immediate income generating resource, and so it was extremely difficult to convince them of the benefits on conservative use of forests. There was no unanimity in decisions because there were fundamental differences regarding the very objectives of the association. The villagers today claim that the neighbouring villages followed their example in forming a forest protection committee and succeeded in doing so, while Maral Mendha, the pioneering village failed in its effort.

A Great Setback

The process of destruction was accentuated in 1988 when construction of Gosikhurd dam on river Wainganga was initiated by the Government of Maharashtra. It is an irrigation project, and a canal to be built under this project is to pass through Maral Mendha. Since, most of the land to be taken over by the government for the canal is forest land, villagers have encroached upon it, hoping to receive compensation money from the government at the time of requisition. The irrigation project is also expected to substantially enhance the prices of agricultural land due to its proximity to the irrigation canal. This too encouraged the residents of Maral Mendha to encroach on the forest land. These encroachments in Maral Mendha do not seem to have taken place out of necessity. They are strategically planned, keeping in mind the possible enhanced future value of land as well as the possibility of being compensated by the government. It can be substantiated from the fact that some of the landowners in Maral Mendha have sold off the small pieces of land that they possessed, and from this money, they have made massive encroachments on the forest land. More than two decades after the declaration of the irrigation project, neither has the land been requisitioned nor has any compensation come the way of Maral Mendha community. The only thing that has happened is total destruction of the forest. Maral Mendha village looks distinctly different from Saigata and Lakhapur. It has concrete houses with tiled roofs, better clothes and more food, larger land holdings to till and more cash in hand. Yet, the women from the village have to travel long distance for fetching fuel wood, men have to buy ploughs and other wooden implements from the nearby market at high prices.

Local indicators of economic state

Local definition of wealth	Own land, big house and possess ion of gold jewellery.
Local definition of poverty	Unemployed and people with insufficient income
Existence of economic disparity based on the local definition	Yes

Failure of Formal Leadership

One wonders at the total passivity of the Forest Department in the whole process. It is difficult to believe that the department officials were unaware of the, not so gradual, destruction of the resource. Apparently, the economically powerful in the village have corrupted the formal owners of the forest. Like in all villages, the village level institution of *gram panchayat* is present in Maral Mendha, but it has no vision regarding sustainable use of forests. Today Maral Mendha is directed by short-term gains that clearly reflect on its state of forest. But the social problems arising out of the situation became visible when our discussions led to heated debates amongst the villagers, indicating clear rift between the rich and the poor, as well as differing perceptions of men and women. Conflicts were clearly on class and gender lines.

Apparently the poor sections of the community are not satisfied with the state of affairs wherein a few have cornered all the gains of illegal encroachment. Women also complained that the men folk think only about increasing their income by acquiring more and more land. They did not consider that there being no forest left for the village use, the women had to collect fuel wood from the neighbouring forests and were subjected to embarrassments if caught. Conflicts also became evident between the landed and the landless. The landless cannot afford to encroach upon forest land due to dearth of resources required to make it cultivable. This class is disadvantaged from both the sides, neither do they have income from agriculture to enable them to buy forest goods, nor are they left with commons to take care of their day-to-day needs of fuel wood and other minor forest products. This class demanded restoration of forest land and asked us to urge the forest department to act firmly against the encroachers and rebuild the forest. They do not have confidence to go and approach the department officials, nor do the officials come to the village to find out the requirements, or to check the state of the "forest".

The community is clear about the internal conflicts and diversity of opinions. It is difficult even to address it as a 'community', which has no shared understanding, no common norms and active forum to get together to share common concerns. It is only a group of individual households living within a village area where each household is trying to maximize its utility irrespective of the impact it has on the whole community. During our discussion one thing the community mentioned was: "we too could have saved our forests only if we had someone like Mr. Suryabhan Khobragade (the leader of Saigata) to lead us and show us the inevitable outcome of our short sighted action". Since there cannot be a 'Suryabhan Khobragade' in each village, the public institutions like forest department or the *gram panchayat*, are expected to provide the necessary leadership and futuristic view. Both these institutions failed miserably in this case of Maral Mendha.

Looking Back in Introspection

The environmental consequences of clearing forest land are visible as no woody species can any longer be found in the forest area, and the land also has become highly degraded. Impact of heavy grazing and damage by pest is visible. High temperature and exposed surface has adversely affected the quality of soil. Sugarcane, a crop that requires lot of water, was grown in the village not very long ago. Sufficient water was available then in the village wells and in the irrigation tank. With cutting of all the trees, soil is not able to retain water, and now the villagers are experiencing lower water levels in the wells, while many wells go dry during summer. For fuel, either the standing trees on individual agricultural fields are cut down, or cow dung that could have been used more productively as manure to agricultural fields, is burnt as fuel. Another significant consequence that the village is now facing, is an increase in the number of insects and pests in their fields. Many villagers accept that it is due to destruction of the forest. "When there was forest, insects got their food from there, where will they go now if not in our fields?" is what they ask.

The failure of the early attempts at conservation and unchecked cutting of the forest, can be attributed to a number of factors that are all endogenous. Lack of effective leadership, typical caste composition with dominance of Maral population that saw cash crops as a source of immediate income and preferred to grow these at the cost of the forests, passing of an irrigation canal through the village increasing prospective prices of agricultural land as well as the likelihood of being compensated by the government once the land was requisitioned seem to be major factors that led the community to destruct its forests. Initially it was the high dependence of neighbouring communities, who did not have their own forest land, which added to the pressure of demand on Maral Mendha's forest. Later large scale and systematic encroachments made by the villagers with an eye only on the short-term gains seem to be the most important factor leading to the present state of affairs in Maral Mendha. The preference of the present over the future has resulted in this "tragedy of commons".

Recent Developments

The villagers do not know how, but the local Member of Legislative Assembly (MLA) recently (July 2002) made some funds available for taking up plantation on a 25-hectare patch in Maral Mendha's forest land. Bramhapuri Division office of Forest Department was entrusted with the execution of the project. It is interesting to note that the forest department did not consult the community regarding choice of species to be planted, nor did the villagers show any interest in the work. Neither did the villagers go for plantation work nor are they guarding the plantations now. For the plantation work, labour from Saigata was employed, who were paid slightly lower than the wages earned by agricultural labour during sowing season. Even the guards employed by the forest department come from other villages. When we specifically asked some of the villagers about how they feel regarding this activity, and how would it help them, the respondents disinterestedly said, "once the trees are big enough, they will be cut down for timber". Even today the community does not wholeheartedly believe in protecting forests for future benefits, thereby discounting future heavily.

Basic details about the village Maral Mendha

Elevation	263 metres ASL
Latitude	20° 37′41.9″ N
Longitude	79°48′26.8″EO
Nearest Town/Market/ Administrative Centre	5 km
Total Population	700
Sex composition	Males (54.29%), Females (45.71%)
Current number of households	125
Nuclear families	67
Literacy	82%
Dominant religions	1.Hindu (72.14%) 2.Mahanubhav (18.57%) 3.Buddhist (9.29%)
Dominant caste groups	1.Mali (50%) 2 Mahar (15%) 3.Teli (10%)

6

BOTANICAL EVIDENCE

It was considered relevant that the botanical health of the forests be considered in order to relatively place the five communities regarding the effectiveness or ineffectiveness of the formal institutions in protection and management of the natural resources. This could prove to be an accurate indicator of the human activities (constructive or destructive) on the state of the forests. Since the quality of the forest can be a good indicator of the success of these management strategies adopted by the communities, we found it appropriate to collect details regarding the health of the forest through forest mensuration, using the IFRI method, as has been mentioned in the first chapter. Accordingly, 30 plots from each village, except Maral Mendha (5 plots) were laid to collect botanical information.

It is a known fact that anthropogenic activities influence various ecological processes, which directly or indirectly affect the processes of regeneration of the plants. Selective logging or habitat fragmentation by forest clearance can result in seed set due to changes in the foraging pattern of poorly mobile pollinators (Ghazoul, 2001). Forest fragmentation in tropics severely affects large trees (Laurence *et al.*, 1997; 1998a; 1998b; 2000; Curran *et al.*, 1999; Gascon *et al.*, 2000), but its effect on other life stages and plant life forms is poorly understood (Benitez-Maldivo, 1998; Scariot, 1999). In the long term, tree species persistence in forest fragments will depend on the availability of seeds, seedlings and saplings. In most cases, species richness declines with fragment size and is independent of seedling abundance (Julieta Benitez-Maldivo and Miguel Martinez-Ramos, 2001). Overall, the life form composition and structure of the regenerated plant pool in fragments shift towards a species-poor seedling community, dominated by lianas, characteristic of disturbed forests. Losses of understorey diversity, especially that of tree seedlings, threaten the maintenance of forest biodiversity and compromise future forest regeneration.

Though all the villages are found to be located on similar land characteristics under uniform climatic conditions, the forest quality registered significant variation. The forests of the region are dominated by species like *Tectona grandis, Albizia odoratissima, Anogeissus latifolia and Terminalia alata*. The physiography of Saigata, Maral Mendha and Lakhapur is more or less flat. Being located in semi-urban periphery, the forests show evidence of more human pressure. Since the three villages share a similar history of denudation of their forests at the hands of first the *malgujars* and then the contractors and poachers, all the density and richness of vegetation that could be found in this cluster of villages, with exception of Maral Mendha, have increased in the past few decades due to the efforts of the community. Lekha and Mendha are on the relatively higher relief, a combination of hills and valleys with occurrence of both upland xeric and bottomland mesic species. The combination of topography, latitude and altitude and also approachability, acutely influence localized composition and structure of the forest close to these villages.

The comparative analysis of the five forests

The forest of Saigata exhibited the positive results of the scientific management practiced by the community. Vegetation of the area is made up of 28 per cent herbs, 45 per cent saplings and 27 per cent trees. The following Diagram 6.1 represents this aspect of the Saigata forest.

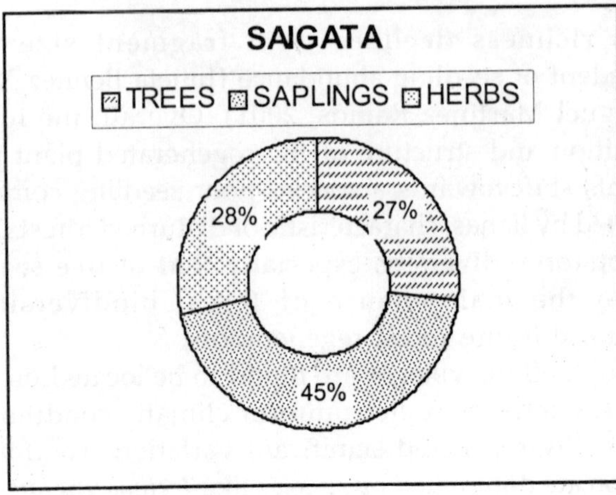

Diagram 6.1

The percentage of saplings being almost double than that of the trees, as evident from Diagram 6.1, is a sign of health in case of the Saigata forest as it implies plantation activities on the part of the community and promises a second generation of trees even when the present ones are harvested. This high density and frequency of saplings and seedlings is the result of protection against poaching and grazing.

The study of the Saigata forest has been undertaken according to the different management zones created by the villagers themselves (see Graph 6.1).

Annual species dominate the *Gawatache beed*. Dominant plant species belong to family Poaceae, Fabaceae and Acanthaceae. The total absence of woody elements is comprehensible as the patch is kept separately for fodder and thatching requirements. However, saplings of *Terminalia* species, *Tectona, Anogeissus latifolia, Cleistanthus, Schleichera oleosa* and *Xeromphis spinosa* form 45 per cent of the vegetation, which suggests that on rotation basis this area can be converted into wood land in near future.

Andhari Van has 60 per cent trees, 29 per cent saplings and 11 per cent herbs and seedlings. Trees are arranged in two strata. Trees of 6-10 metres height and 15-19 cms DBH (Diameter at Breast Height) occupy the top stratum. A second

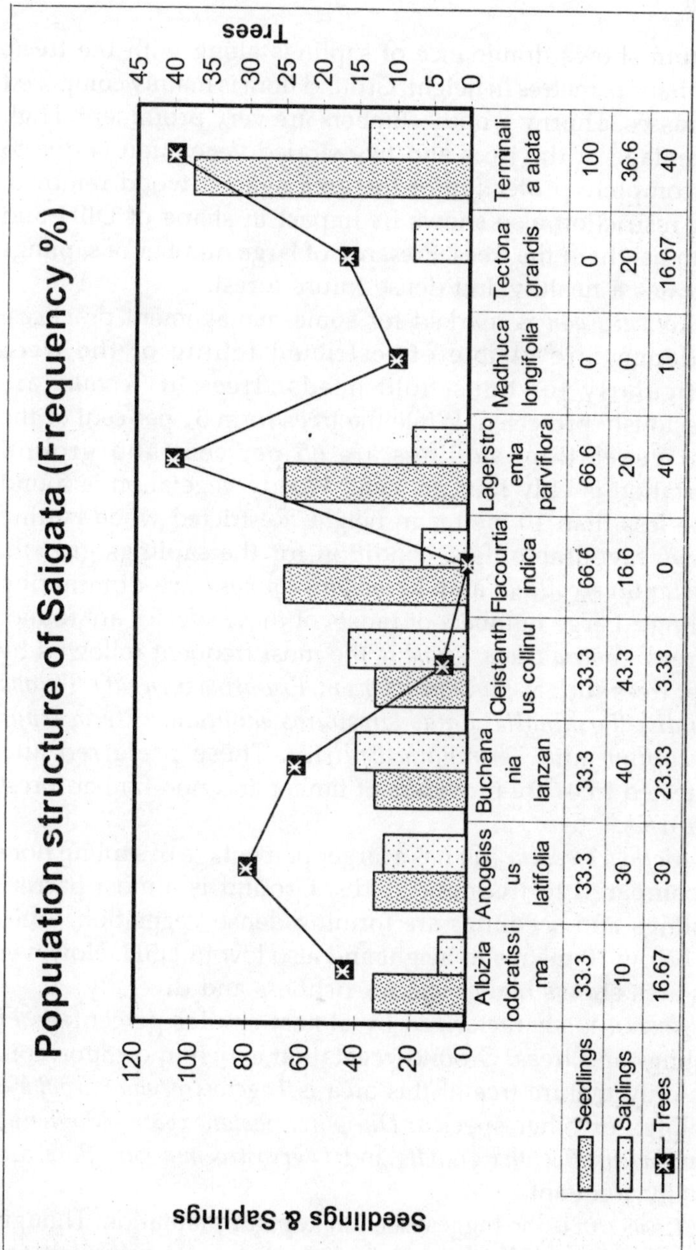

	Albizia odoratissima	Anogeissus latifolia	Buchanania lanzan	Cleistanthus collinus	Flacourtia indica	Lagerstroemia parviflora	Madhuca longifolia	Tectona grandis	Terminalia alata
Seedlings	33.3	33.3	33.3	33.3	66.6	66.6	0	0	100
Saplings	10	30	40	43.3	16.6	20	0	20	36.6
Trees	16.67	30	23.33	3.33	0	40	10	16.67	40

Graph 6.1

stratum shows dominance of saplings along with the trees, less than six metres in height. Ground flora is mainly composed of grasses. Thorny woody climbers are very prominent. High percentage of the trees and two-storied vegetation is due to the complete protection of the area against wood removal. This restriction also shows its impact in shape of DBH and mean height of the trees. Presence of large number of saplings indicates a healthy and dense future forest.

Kotwal Raan is marked for some management practices. Evidences are visible of restricted felling of the trees particularly for household needs. Trees of *Tectona* are consciously protected. While the trees form 31 per cent of the total vegetation, saplings are 65 per cent and ground vegetation is only four per cent. Woody vegetation is found to be less than 10 metres in height. Restricted wood-cutting allows favourable light condition for the saplings growth. *Cleistanthus collnus* and *Xeromphis spinosa* are dominating saplings. Large numbers of ranets of these species are found. In trees, *Albizia odoratissima* is the most frequent followed by other trees such as *Buchanani lanzan, Erythrina veriegata, Tectona grandis, Terminalia alata, Diospyros montana, Pterocarpus marsupium* and *Terminalia bellirica*. These preferred and protected trees are providers of timber and non-timber forest produces.

Mothya Devache van has a larger percentage of sapling flora as compared to trees and herbs. Ground is almost barren. Saplings and seedlings are forming dense vegetation. Trees are below 10 metres in height and also low in DBH. However this area shows higher species richness and diversity.

Kosran is characterized by almost similar percentage of saplings and trees. Ground vegetation is rich in composition. The only mature tree of this area is *Lagerostroemia parviflora*. Saplings of other species, *Diospyros melanoxylon, Maytenus emarginatus, Tectona grandis* and *Lagerostroemia parviflora* are equally frequent.

Tiwas van is the biggest section kept for plantation. Though the trees are small, the area is rich in species composition. *Anogeisus latifolia, Buchanania lanzan* and *Cleistanthus collinus* species form dominating saplings, while *Terminalia alata*,

Anogeisus, Lagerostroemia parviflora and *Buchanania lanzan* are more prominent trees.

Analysis of the vegetation of the different management zones of Saigata does not show much variation in the height and DBH of the trees. This may be because of the same age of the vegetation. However variation lies in the ratio of herb, shrub trees and between seedling, sapling and mature trees. The areas that are under complete protection are rich in trees. Those under selective cutting along with other management practices like plantation, are rich in saplings and some are kept strictly for grazing purpose. This forest is in the early stage of succession. Balance in the seedling, sapling and tree ratio predicts a bright future for the forest.

Lakhapur vegetation shows almost equal percentage of saplings and trees, while annuals are forming more than half the vegetation. Trees of Lakhapur are small in height and DBH. Seedlings are least in number, followed by saplings and then trees, which point to the absence of any plantation activity. Regeneration activity in the area is hampered by more anthropogenic influence, less protection and absence of plantation activities. The Importance Value Index of saplings is less than the trees, it predicts a bleak future for the Lakhapur forest. Absence of the *Tectona* is surprising, since it is a precious

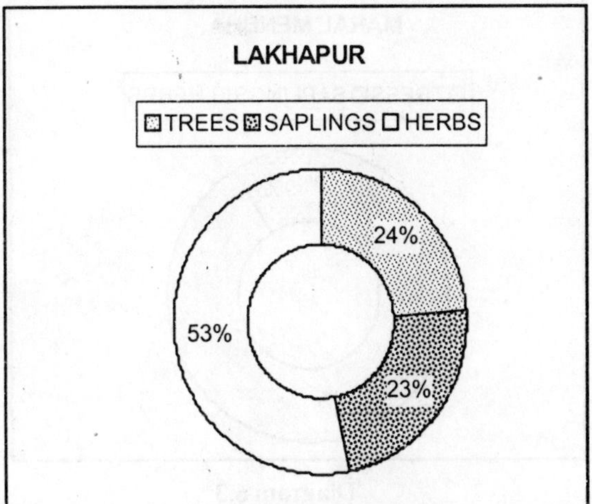

Diagram 6.2

as well as a naturally dominant species of the region. Similarly absence of the some of the important timber yielding species like, *Albizia odoratissima, Anogeissus latifolia, Flacourtia indica, Lagerstroemia parviflora* and *Madhuca longifolia* points at cutting of the trees and disturbed habitat.

The case of Lakhapur is unique in the way that in spite of practicing restrictive use and undertaking other conservation activities, the forest does not seem to have a future in the absence of funds required for the rejuvination and upkeep of the forest. In Lakhapur absence of seedlings of *Albizia* species, *Anogeissus latifolia, Flacourtia indica, Lagerstroemia parviflora, Madhuca longifolia* and *Tectona grandis*, and less number of the saplings is due to the over harvesting (see Graph 6.2). Lack of plantation activities reflects in the fact that the percentage of herbs exceeds both trees and saplings, and that the percentage of saplings is less than that of the trees.

Maral Mendha cannot be compared to the rest of the villages. As mentioned earlier, there is no forest worth talking about in Maral Mendha. The two hectares of land that is left unencroached out of the substantial 71.46 hectares of forest land, is devoid of even a single tree. Herbs constitute almost all of the vegetation, and saplings form a meagre nine per cent.

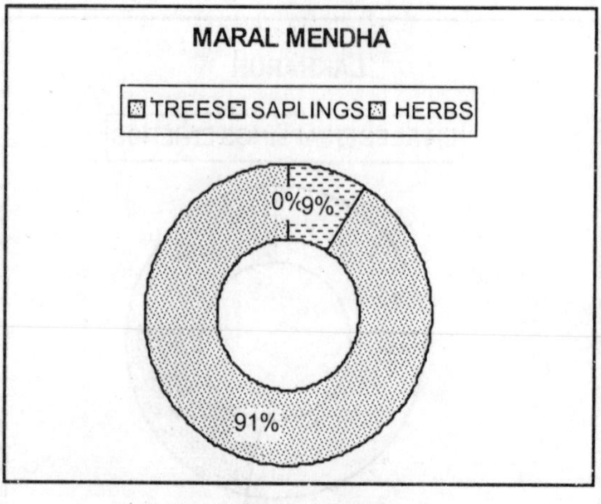

Diagram 6.3

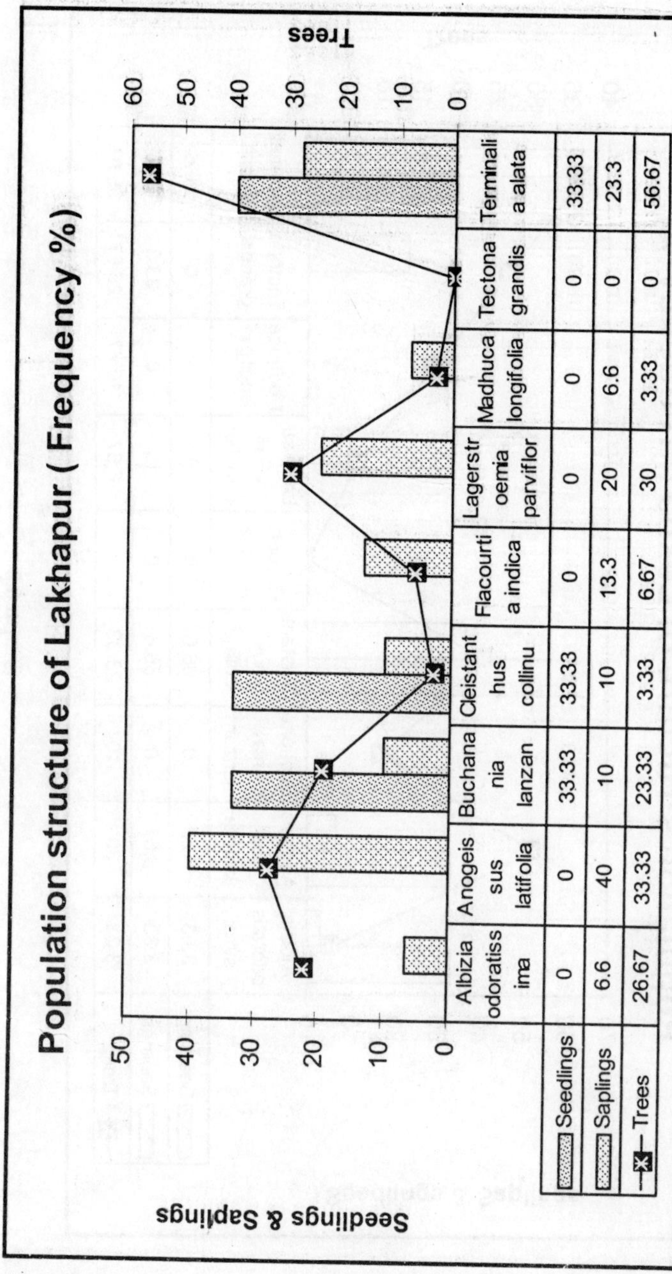

	Albizia odoratissima	Anogeissus latifolia	Buchanania lanzan	Cleistanthus collinu	Flacourtia indica	Lagerstroemia parviflor	Madhuca longifolia	Tectona grandis	Terminalia alata
Seedlings	0	0	33.33	33.33	0	0	0	0	33.33
Saplings	6.6	40	10	10	13.3	20	6.6	0	23.3
Trees	26.67	33.33	23.33	3.33	6.67	30	3.33	0	56.67

Graph 6.2

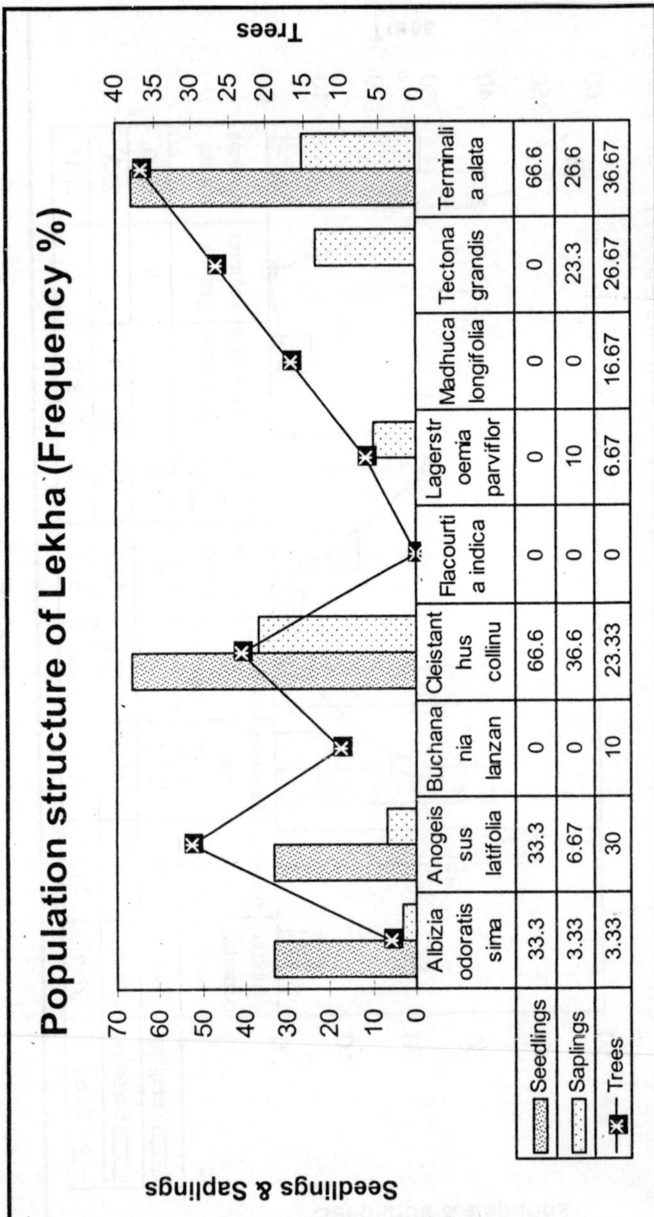
Population structure of Lekha (Frequency %)
Seedlings & Saplings
Trees
Seedlings
Saplings
Trees
Albizia odoratis sima
Anogeis sus latifolia
Buchana nia lanzan
Cleistant hus collinu
Flacourti a indica
Lagerstr oemia parviflor
Madhuca longifolia
Tectona grandis
Terminali a alata
Seedlings 33.3 33.3 0 66.6 0 0 0 0 66.6
Saplings 3.33 6.67 0 36.6 0 10 0 23.3 26.6
Trees 3.33 30 10 23.33 0 6.67 16.67 26.67 36.67

Graph 6.3

Ninty one per cent of the vegetation found in this 2-hectare patch happens to be of annual nature. While saplings are miniscule in number, seedlings are totally absent. This patch is incapable of yielding anything beyond some fodder for the cattle. Forest mensuration data reveals that the total absence of tree cover and high temperature is responsible for lowering the soil moisture regime, which in turn is likely to affect the crop output in the near future.

Lekha forest was found to be composed of 47 per cent herb, 21 per cent saplings and just 32 per cent trees, as shown in Diagram 6.4 below.

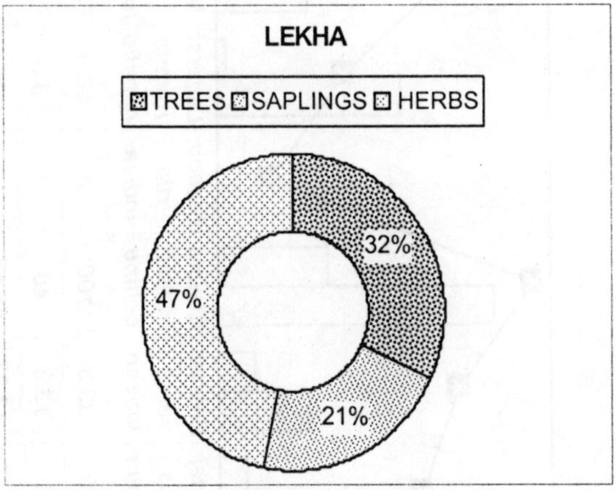

Diagram 6.4

Though, some of the trees are the oldest trees of the area, the part of the forest, which comes under the flat land, has been encroached upon and converted into agricultural land. *Terminalia alata* along with the *Anogeissus latifolia* and *Tectona grandis* form the dominating timber trees of the area. Population structure of the trees shows absence of the seedlings of the *Buchanania lanzan, Lagerstroemia parviflora, Madhuca longifolia* and *Tectona grandis* (See Graph 6.3). Unregulated extraction of the non-timber forest produces [NTFP] is the main reason behind this absence.

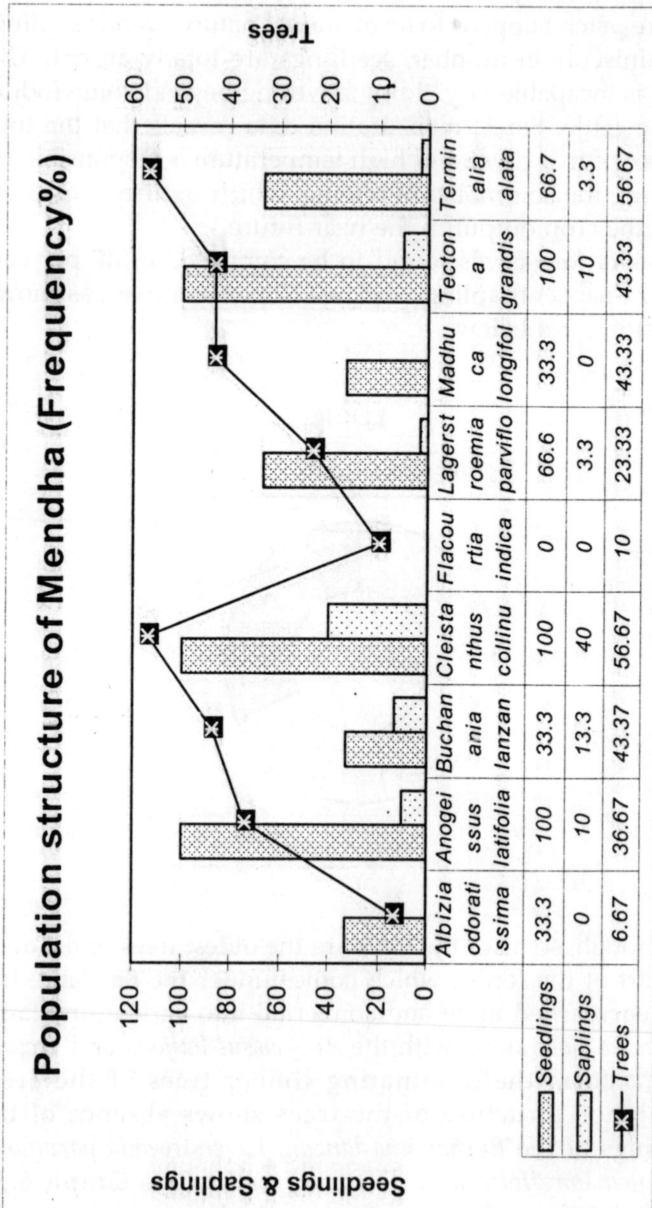
Population structure of Mendha (Frequency%)
Seedlings & Saplings
Trees
120
100
80
60
40
20
0
60
50
40
30
20
10
0
Albizia odoratissima
Anogeissus latifolia
Buchanania lanzan
Cleistanthus collinu
Flacourtia indica
Lagerstroemia parviflo
Madhuca longifol
Tectona grandis
Terminalia alata
Seedlings 33.3 100 33.3 100 0 66.6 33.3 100 66.7
Saplings 0 10 13.3 40 0 3.3 0 10 3.3
Trees 6.67 36.67 43.37 56.67 10 23.33 43.33 43.33 56.67

Graph 6.4

Mean DBH of Trees (in cms)

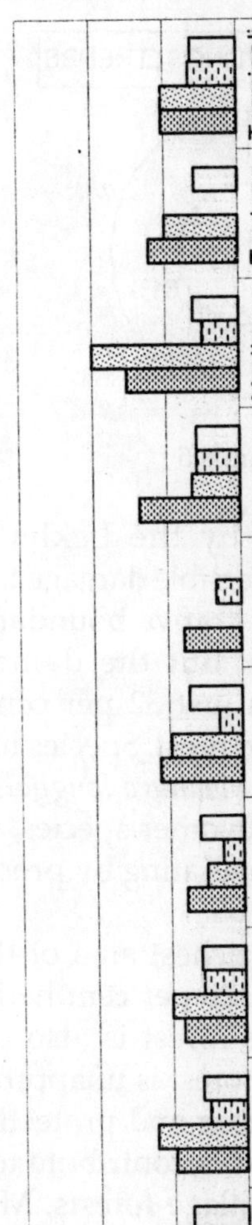

	Albizia odoratissus	Anogeissus	Buchanania	Cleistanthus	Erythrina veriegata	Flacourtia indica	Lagerstroemia	Madhuca longifolia	Tectona grandis	Terminalia alata
Mendha	29.4	19.3	17.1	15.4	22.8	15.5	27.5	30.7	24.2	20.6
Lekha	19.3	24.4	20.1	9.8	19.3	0	13	40.3	19.8	20.6
Lakhapur	11.9	9.9	11.2	5.8	6.3	6.7	11.9	9.8	0	13.2
Saigata	14.93	11.76	11.65	12	14.45	0	11.74	12.2	11.75	12.65

Graph 6.5

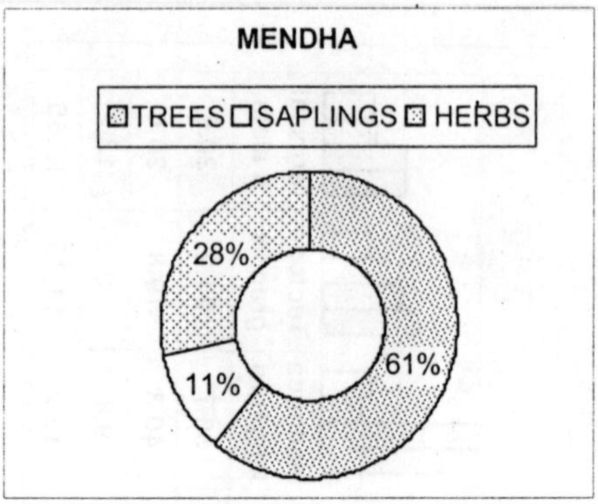

Diagram 6.5

The reckless harvesting by the Lekha community, however, could not do any irreversible damage to the resource, as the forest within the administrative boundary of this area was initially very abundant. But the damage was still considerable as the trees form just 32 per cent of the total vegetation of the once abundant forest. Species like *Buchanania lanzan, Lagerstroemia parviflora, Madhuca longifolia* and *Tectona grandis* have no seedlings, while other species, which have a few seedlings, are actually regenerating by producing ranets. (See Graph 6.3)

Mendha is the floristically richest area of this region. 61 per cent of the plants are trees, 28 per cent herbs and 11 per cent saplings and shrubs. This forest is also the biggest in dimension. A major part of the forest is unapproachable. Less population pressure, tough terrain and protection measures, along with the afforestation efforts contribute to the richness of the forest. Out of the four village forests, Mendha shows highest values for frequency, density and abundance of the trees. Similarly, higher DBH and height of the trees are indicators of effective protection of the vegetation. Trees form a considerable portion of the total vegetation, however saplings are showing comparatively lesser values on the same count. This may be because of closer crown canopy and unfavourable conditions. The tree, seedling and sapling composition of the Mendha forest is graphically represented above.

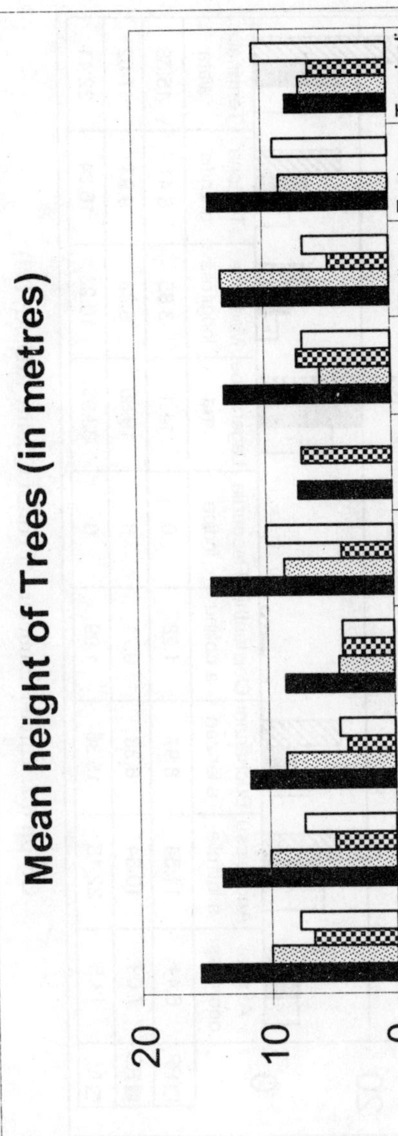

	Albizia odoratis	Anogeissus	Buchanania	Cleistanthus	Erythrina veriegat	Flacourtia indica	Lagerstroemia	Madhuca longifolia	Tectona grandis	Terminalia alata
Mendha	15.5	13.7	11.5	8.6	14.3	7.4	13.1	13.1	14.2	7.9
Lekha	9.9	9.9	8.6	4.3	8.5	0	5.6	13.2	8.6	6.9
Lakhapur	6.6	4.8	3.8	4	4.1	7.1	7.4	4.8	0	6.2
Saigata	7.66	7.26	4.37	4	10	0	6.9	6.7	8.95	10.5

Graph 6.6

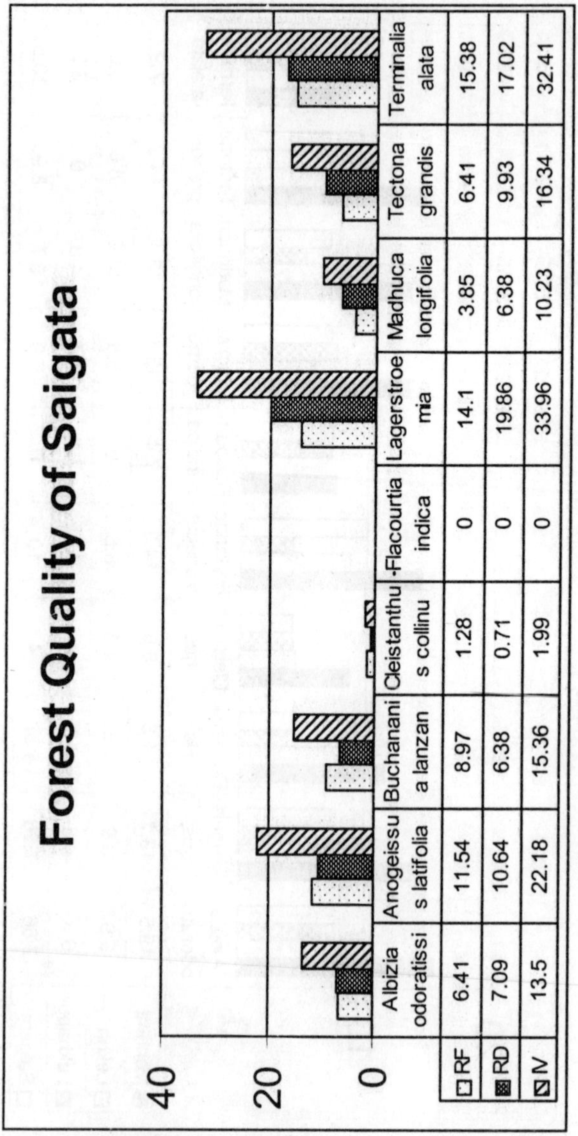

	Albizia odoratissi	Anogeissus latifolia	Buchanania lanzan	Cleistanthus collinu	Flacourtia indica	Lagerstroemia	Madhuca longifolia	Tectona grandis	Terminalia alata
RF	6.41	11.54	8.97	1.28	0	14.1	3.85	6.41	15.38
RD	7.09	10.64	6.38	0.71	0	19.86	6.38	9.93	17.02
IV	13.5	22.18	15.36	1.99	0	33.96	10.23	16.34	32.41

Graph 6.7: Three Parameters of Forest Quality (RD, RF, IVI)

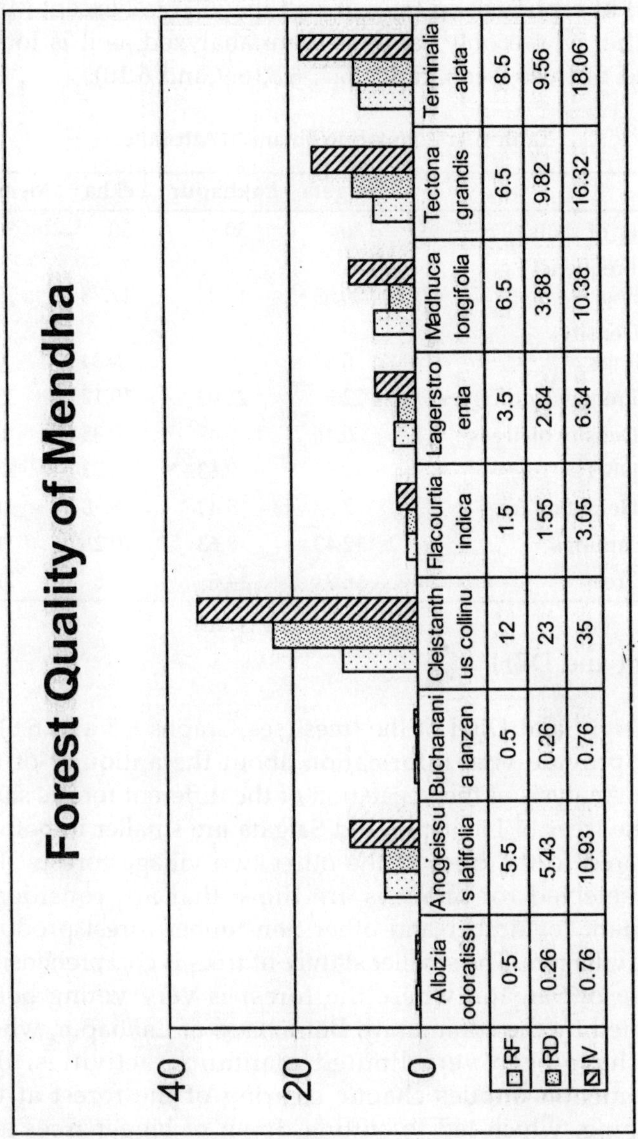

Graph 6.8: Three Parameters of Forest Quality (RD, RF, IVI)

Table 6.1 presents some prominent botanical attributes like mean frequency, DBH, height, and IVI (Importance Value Index) of saplings and trees, for all the villages, except Maral Mendha, where only saplings were analyzed as it is totally devoid of trees (see Graphs 6.7, 6.8, 6.9 and 6.10).

Table 6.1: Comparing Botanical Attributes

	Saigata	Lakhapur	Lekha	Mendha
Number of Plots	30	30	30	30
Mean Frequency of Sapling	30.83	13.34	17.78	13.33
Mean Density of Saplings	0.87	0.26	0.34	0.37
Mean Frequency of Trees	22.5	22.92	19.17	35.5
Mean Density of Trees	0.46	0.47	0.35	0.88
Mean DBH of Trees	12.57	9.63	20.73	22.25
Mean Height of Trees	7.37	5.42	8.38	11.93
IVI of Saplings	12.43	8.63	10.28	11.7
IVI of Trees	16.21	14.5	9.8	11.28

Height and DBH

The height and DBH of the trees (see Graphs 6.5 and 6.6) in forest provide vital information about the antiquity of the forest. Analysis of the vegetation of the different forests show that the trees of Lakhapur and Saigata are smaller in both as compared to the trees of the other two village forests. The trees selected for analysis are those that are considered important for timber and other non-timber forest products by the villagers. The smaller stature of trees is comprehensible in case of Saigata, where the forest is very young being replanted by the community. But in case of Lakhapur, where there have been very limited plantation activities, this phenomenon implies chaotic clearing of the forest at the beginning, which led to cutting down of bigger trees and also ineffective protection. While comparing Lakhapur with

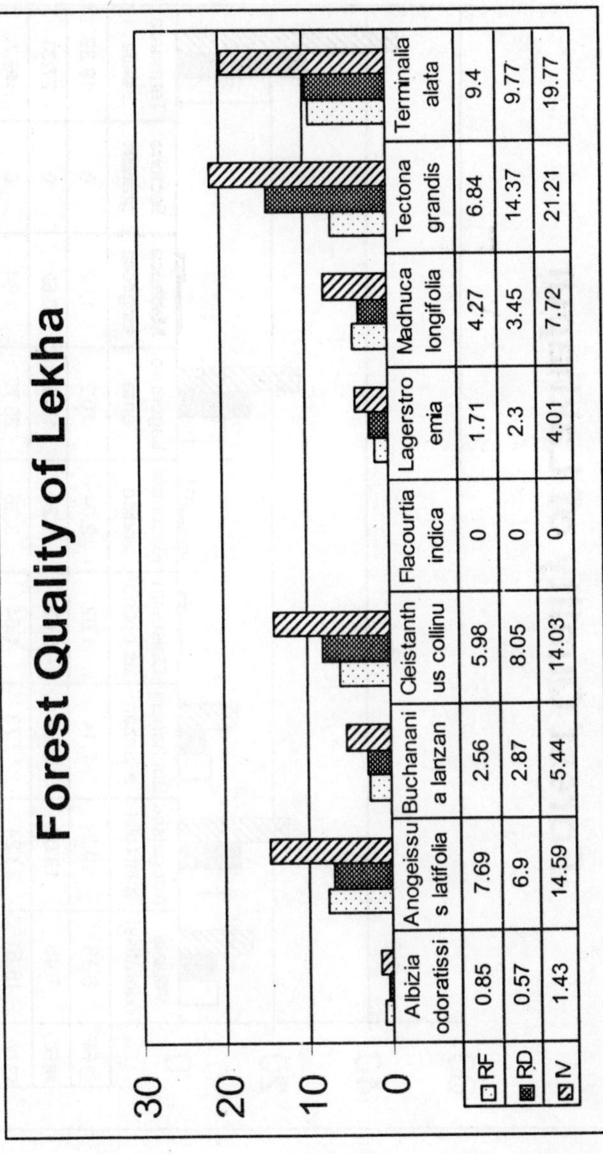

	Albizia odoratissi	Anogeissu s latifolia	Buchanani a lanzan	Cleistanth us collinu	Flacourtia indica	Lagerstro emia	Madhuca longifolia	Tectona grandis	Terminalia alata
RF	0.85	7.69	2.56	5.98	0	1.71	4.27	6.84	9.4
RD	0.57	6.9	2.87	8.05	0	2.3	3.45	14.37	9.77
IVI	1.43	14.59	5.44	14.03	0	4.01	7.72	21.21	19.77

Graph 6.9: Three Parameters of Forest Quality (RD, RF, IVI)

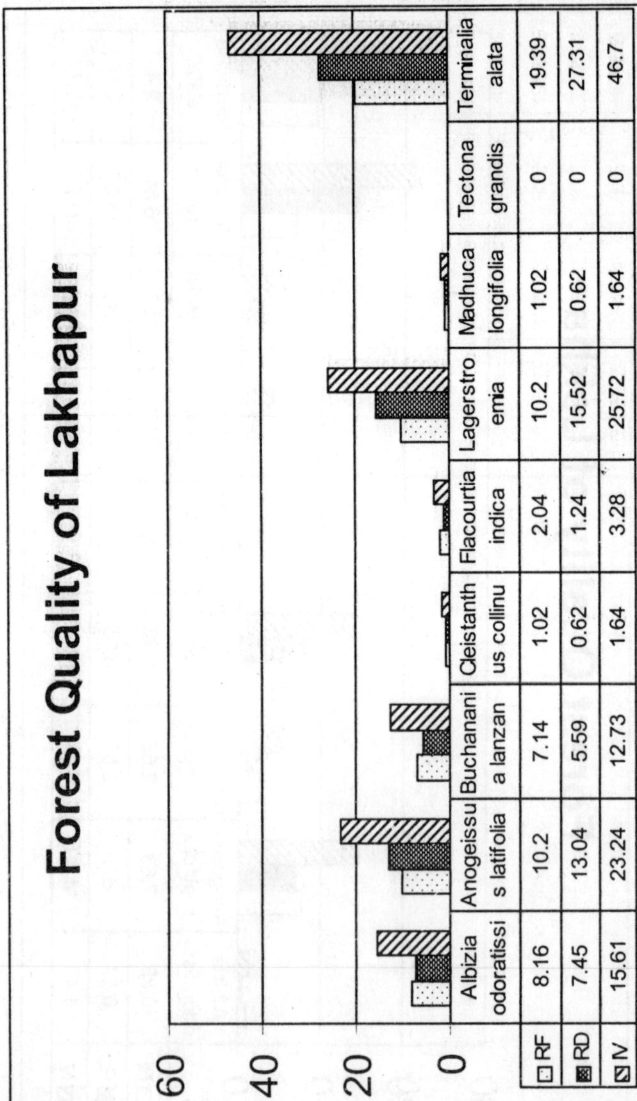

	Albizia odoratissi	Anogeissus latifolia	Buchanania lanzan	Cleistanthus collinu	Flacourtia indica	Lagerstroemia	Madhuca longifolia	Tectona grandis	Terminalia alata
RF	8.16	10.2	7.14	1.02	2.04	10.2	1.02	0	19.39
RD	7.45	13.04	5.59	0.62	1.24	15.52	0.62	0	27.31
IV	15.61	23.24	12.73	1.64	3.28	25.72	1.64	0	46.7

Graph 6.10: Three Parameters of Forest Quality (RD, RF, IVI)

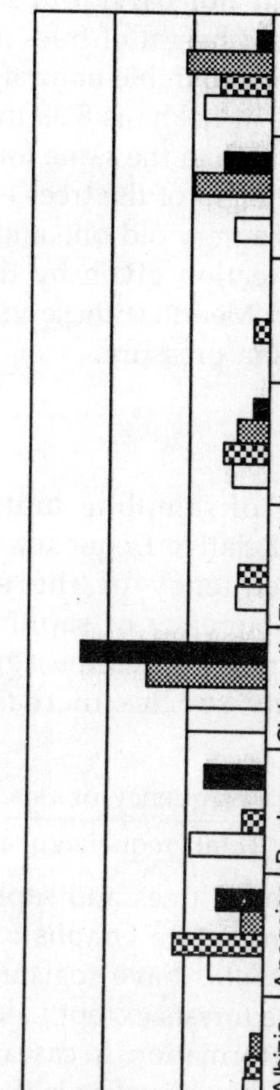

	Albizia odoratissi	Anogeissus latifolia	Buchanania	Cleistanthus collinu	Flacourtia indica	Lagerstroemia	Madhuca longifolia	Tectona grandis	Terminalia alata
□ Saigata	2.33	6.98	9.3	10.08	3.88	4.65	0	4.65	8.53
⊠ Lakhapur	1.9	11.43	2.86	0	3.81	5.71	1.9	0	6.67
▩ Lekha	1.37	2.74	0	15.07	0	4.11	0	9.59	10.96
■ Mendha	0	5.88	7.84	23.53	0	1.96	0	5.88	1.96

Graph 6.11

Saigata, it is clearly visible that the mean height and DBH of trees of Saigata are more than those of Lakhapur.

Though the Lekha community could not protect its forest as desired and also could not carry out any plantation activities, the mean DBH and height of trees in the forest are not very low mainly due to favourable natural characteristics. Average height of the trees in Lekha is 8.38 metres and DBH is 20.73 cms, which are more than the same for Lakhapur and Saigata. Higher DBH and height of the trees exist in Mendha as the forest of this area is a very old one and this also is an indicator of effective protection given by the community. However, both (Lekha and Mendha) these villages have the advantage of low population pressure.

Frequency

Frequency is the number of sampling units in which the particular species occurs. Relative frequency of saplings can be described as the frequency of the sapling under consideration over the frequency of saplings of all other species. Relative frequency indicates the aggressiveness of the species by which any species increases its area of occurrence.

$$\text{Relative Frequency} = \frac{\text{Frequency of a specie} \times 100}{\text{Total Frequency of all species}}$$

The relative frequencies of trees and saplings have been worked in case of four forests (see Graphs 6.11 & 6.12). The saplings of *Cleistanthus collinus* have registered the highest relative frequency in all the forests except Lakhapur. This has happened because of ranet formation. In case of trees however the relative frequency of *Terminalia alata* is the highest for all the four villages. Lower relative frequencies of *Tectona grandis, Buchanania lanzan,* and *Madhuca longifolia* saplings are also surprising as all the three species are of economic significance. While the first is a precious timber yielding specie, the next two are producers of economically important non-timber-forest-products. For timber, the communities usually use

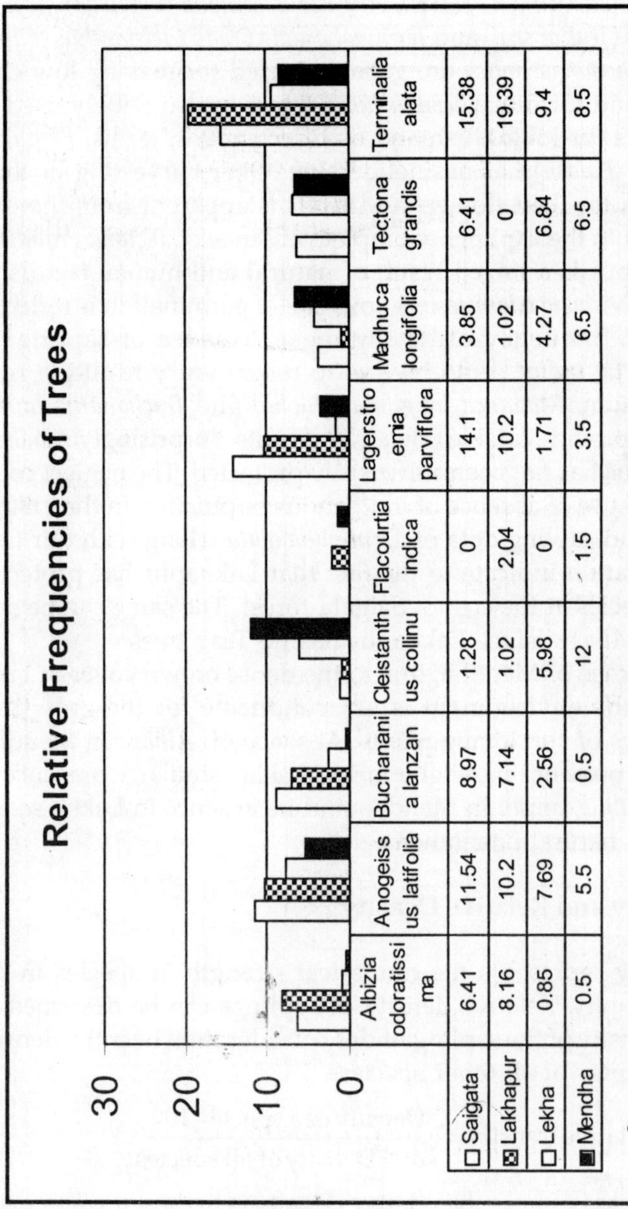

	Albizia odoratissima	Anogeissus latifolia	Buchanania lanzan	Cleistanthus collinus	Flacourtia indica	Lagerstroemia parviflora	Madhuca longifolia	Tectona grandis	Terminalia alata
Saigata	6.41	11.54	8.97	1.28	0	14.1	3.85	6.41	15.38
Lakhapur	8.16	10.2	7.14	1.02	2.04	10.2	1.02	0	19.39
Lekha	0.85	7.69	2.56	5.98	0	1.71	4.27	6.84	9.4
Mendha	0.5	5.5	0.5	12	1.5	3.5	6.5	6.5	8.5

Graph 6.12

Albizia odoratissima (Siris), Angeissus latifolia (Dhawda), Tectona grandis (Teak or saal) and *Terminalia alata (Yen). Cleistanthus collinus* and *Flacourtia indica* are generally used for making low class tools and for fuel. *Cleistanthus collinus* is also widely used by farmers for killing fish and as insecticide in farms. *Erythrina* (*Ghada Palash*) has aesthetic value. Over harvesting of these species for their flowers and fruits is apparent from the data related to the saplings. Complete absence of *Tectona grandis* in Lakhapur is a mixed result of natural and human factors. In case of *Albizia odoratissima,* low rate of germination is reflected in low frequency of its saplings. Presence of saplings of *Flacourtia indica* could be due to recent entry resulting from protection. *Madhuca longifolia* (*Moha)* and *Buchanania lanzan (Char)* provide eatable flowers and fruits. Surprisingly, *Madhuca longifolia* has not been sufficiently protected. The neglect could be because of absence of indigenous population in the village. Tree and sapling data of *Terminalia alata,* along with our field observation indicate to the fact that Lakhapur has protected the species at the cost of Saigata forest. The same can be said about *Albizia,* since Lakhapur has no Teak trees.

In case of Mendha, due to the dense crown cover of taller trees, the environment is not conducive for the growth of saplings of sun loving plants. Absence of *Albizia* in Mendha, but its presence in neighbouring Lekha, similarly presence of *Buchanania lanzan* in Mendha and its absence in Lekha seems to be a natural phenomena.

Density and Relative Density

Density represents the numerical strength of species in the community. Relative density of saplings can be described as the density of the sapling under consideration over the density of saplings of all other species.

$$\text{Relative Density} = \frac{\text{Density of a specie} \times 100}{\text{Total Density of all species}}$$

In case of trees the relative density is highest for *Terminalia alata*. It is found to be 27.31 in Lakhapur and 17.12 in Saigata.

It is considerably high in the other two forests also. The relative density of *Flacourtia indica* trees is uniformly low in Lakhapur and Mendha forests and is non-existent in the forests of Lekha and Saigata (see graphs). In case of saplings, however, *Madhuca longifolia* has the lowest relative density. Its saplings were sighted only in the forest of Lakhapur.

It is disheartening to note that even in case of Saigata and Mendha, where protection effort is more than Lakhapur and Lekha, the relative density of trees is much more than relative density of saplings. It also indicates the urgent need of positive intervention in forest management in the form of *in-situ* plantation. This is true about all the species except *Cleistanthus collinus* (see Graphs 6.13 and 6.14).

Importance Value Index (IVI)

Importance Value Index is the sum total of relative frequency and relative dominance. It is an indicator of aggressiveness, present status of the species and also of the future prospect of the forest.

Importance Value Index = Relative Frequency + Relative Density

Dominance is calculated as the basis of IVI of the species. It is highest in Saigata, followed by Mendha, Lakhapur and Lekha. In our sample, in case of IVI of trees, it is clear that *Terminalia, Cleistanthus* and *Lagerstroemia* have higher value. Of these the *Cleistanthus* and *Lagerstroemia* are reproduced by sprouting. Hence they do not depend upon other contravines for seed formation. *Anogeisus* and *Tectona* also rank high on value index. *Terminalia* and *Anogeisus* have higher value in Lakhapur in comparison with Saigata. *Buchanania lanzan, Lagerstroemia parviflora* and *Tectona grandis* have higher importance value in Saigata. While comparing Lekha and Mendha, *Lagerstroemia parviflora, Madhuca longifolia* and *Cleistanthus* have more value in Mendha than in Lekha. Lekha has more value for *Buchanania lanzan, Tectona grandis* and *Anogeisus latifolia*. Similar trends can be seen in IVI of saplings as well (See Graphs 6.15 and 6.16).

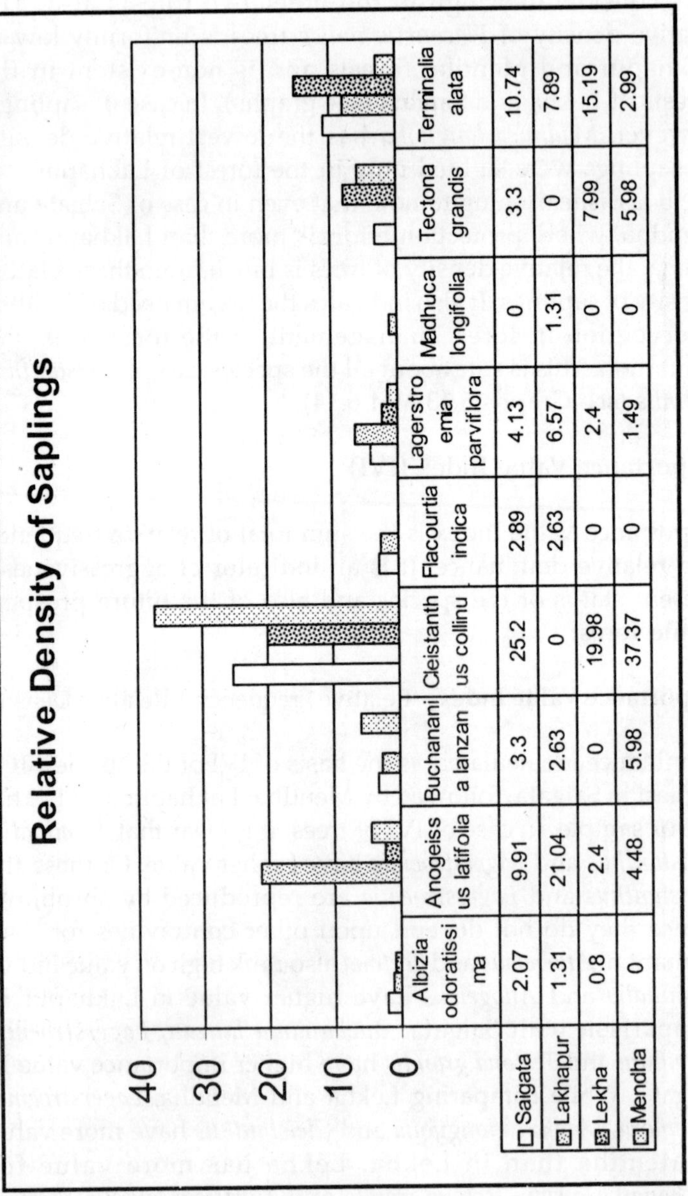

	Albizia odoratissima	Anogeissus latifolia	Buchanania lanzan	Cleistanthus collinus	Flacourtia indica	Lagerstroemia parviflora	Madhuca longifolia	Tectona grandis	Terminalia alata
Saigata	2.07	9.91	3.3	25.2	2.89	4.13	0	3.3	10.74
Lakhapur	1.31	21.04	2.63	0	2.63	6.57	1.31	0	7.89
Lekha	0.8	2.4	0	19.98	0	2.4	0	7.99	15.19
Mendha	0	4.48	5.98	37.37	0	1.49	0	5.98	2.99

Graph 6.13

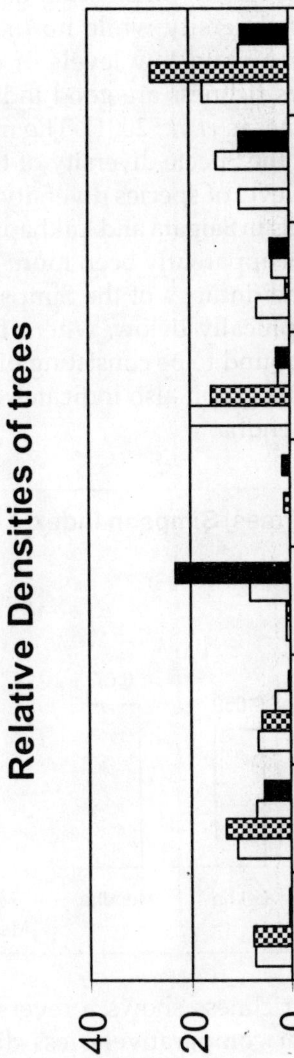

	Albizia odoratissi	Anogeiss us latifolia	Buchanan ia lanzan	Cleistanth us collinu	Flacourtia indica	Lagerstro emia	Madhuca longifolia	Tectona grandis	Terminalia alata
Saigatha	7.09	10.64	6.38	0.71	0	19.86	6.38	9.93	17.02
Lakhapur	7.45	13.04	5.59	0.62	1.24	15.52	0.62	0	27.31
Lekha	0.57	6.9	2.87	8.05	0	2.3	3.45	14.37	9.77
Mendha	0.26	5.43	0.26	23	1.55	2.84	3.88	9.82	9.56

Graph 6. 14

Simpson and Shannon Indices

Forest change and stability may have important implications for the maintenance of biodiversity. Small-scale disturbance may lead to enrichment of diversity while no or minimal disturbance may actually maintain low levels of diversity. Species diversity and species richness are good indicators of the level of disturbance [Ganesan *et al.*, 2001]. The number of species of trees in a forest is the Specie diversity of the forest. Simpson Index exhibits low level of species diversity in Lekha and Mendha, but higher levels in Saigata and Lakhapur, where the extent of disturbance has apparently been more than that in the other two villages. The findings of the Simpson Index have been represented graphically below, where forests of Saigata and Lakhapur were found to be consisting of a highly diversified range of trees. The graph also indicates complete absence of trees in Maral Mendha.

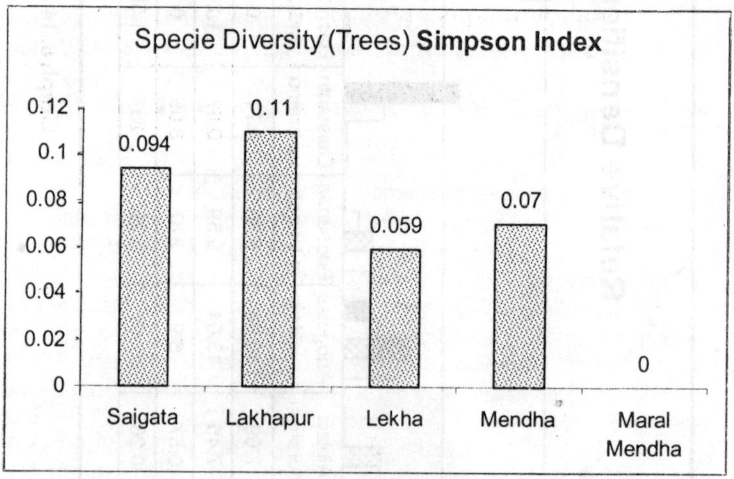

On the contrary species richness shows a reverse trend. It indicates higher values for comparatively less disturbed Lekha and Mendha, and lower value for other two village forests.

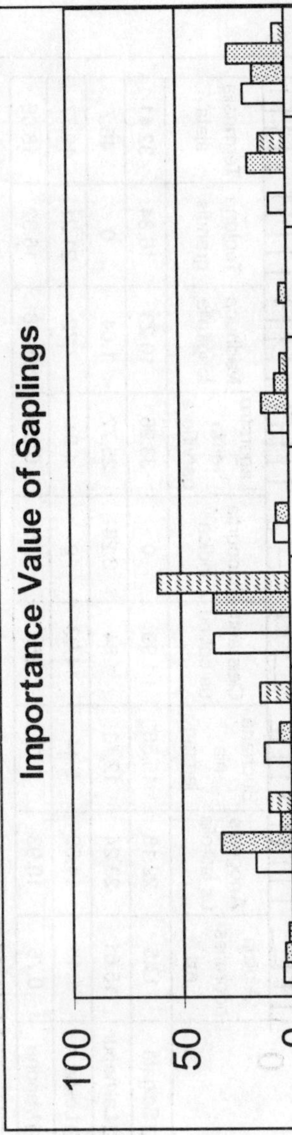

	Albizia odoratissi	Anogeiss us latifolia	Buchanani a lanzan	Cleistanth us collinu	Flacourtia indica	Lagerstro emia	Madhuca longifolia	Tectona grandis	Terminalia alata
Saigata	4.39	16.89	12.61	35.27	6.77	8.78	0	7.96	19.27
Lakhapur	3.22	32.47	5.49	0	6.44	12.29	3.22	0	14.56
Lekha	2.17	5.14	0	35.05	0	6.51	0	17.58	26.15
Mendha	0	10.37	13.82	60.9	0	3.46	0	11.86	4.95

Graph 6.15

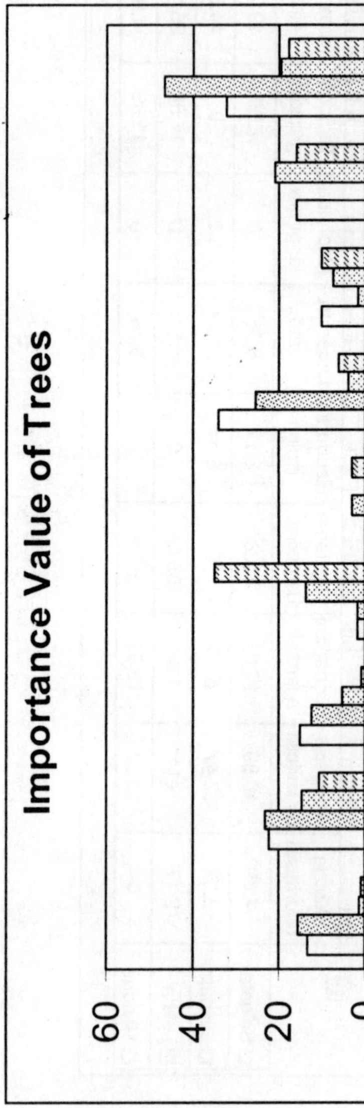

	Albizia odoratissima	Anogeissus latifolia	Buchanania lanzan	Cleistanthus collinus	Flacourtia indica	Lagerstroemia parviflora	Madhuca longifolia	Tectona grandis	Terminalia alata
Saigata	13.5	22.18	15.36	1.99	0	33.96	10.23	16.34	32.41
Lakhapur	15.61	23.24	12.73	1.64	3.28	25.72	1.64	0	46.7
Lekha	1.43	14.59	5.44	14.03	0	4.01	7.72	21.21	19.77
Mendha	0.76	10.93	0.76	35	3.05	6.34	10.38	16.32	18.06

Graph 6.16

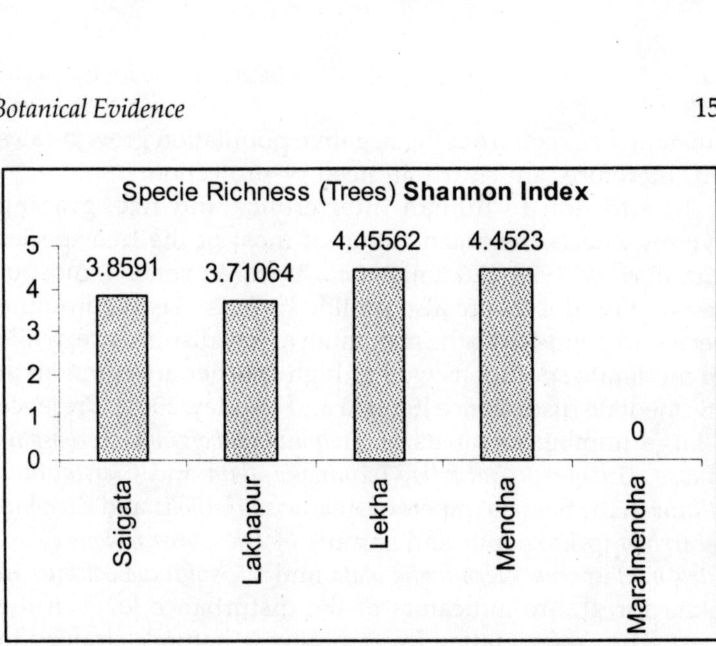

Extraction of various non-timber forest produces and other human induced disturbances such as fire affect the population dynamics and demography of the harvested species, and ultimately the sustainability of extraction [Sinha and Bawa, 2001, Daniels *et al.*, 1995]. Removal of the fruits from the forest has a smaller effect on population growth rate of *Phyllanthus* spp. than the existing fire regimes. However, the current fruit harvesting techniques such as lopping off of the entire canopy and cutting of the trees have negative impact on *Phyllanthus* trees. Disturbance has been considered as an important regulator of community structure [Foster, 1980], but its quantification is the major problem for the study of its relationship with community phenomenon [Huston, 1994]. Survival of reproductive adults is an important parameter that determines the rate at which the population grows. In Lekha, *Buchanania lanzan*, *Lagerstroemia parvifolia* and *Madhuca longifolia* are good examples of the over-exploitation of the plants as seedlings and saplings are absent. Same phenomenon of unjust exploitation is observed in the Lakhapur forest. Local populations *of Albizia odoratissima, Anogeissus latifolia, Madhuca longifolia* and *Lagerstroemia parviflora* are threatened as seen from the negative population growth rates and, therefore, are clearly in need of protection.

In addition to human interference and fire, grazing severely affects the regeneration of most of the tree species [Murali *et al.*, 1996, Sukumar *et al.*, 1998]. In general, most of the ranet producers are also prolific sprouts. Least sprouting species are found to be the maximum at low disturbance, while the moderate sportier as well as high sportier are common at intermediate disturbance [Shukla and Pandey, 2001]. Presence of large number of ranets of *Ziziphus glaberrima, Xeromphis spinosa, Anogeissus latifolia, Terminalia alata* and *Cleistanthus collinus* in the Saigata, ranets of *Anogeissus latifolia* and *Ziziphus glaberrima* in Lakhapur and sprouts of *Diospyros melanoxylon, Butea monosperma, Terminalia alata* and *Cleistanthus collinus* in Lekha forest, are indicators of the disturbance level in the forest. Thus regeneration by sprouting is nature's strategy to withstand the vagaries of high disturbance.

The impacts of selective logging on tree regeneration may be positive or negative and may sometimes be neutral [Horne and Hickey, 1991; Guariguata & Pinnard, 1998; Putz *et al.*, 2000; Pinard & Swaine, 2001]. The principal patterns identified are: [1] where soil is damaged, tree generation and growth are restricted as in Maral Mendha; [2] where felling gap size exceeds a certain threshold, tree growth and survival are reduced by competition with weedy vegetation [e.g. Lianas, pioneer vegetation], as in Lekha; [3] Early succession species are likely to become more abundant over space and time unless cutting cycles are long as in Saigata; and [4] successful regeneration may be more likely where there was an existing seedling bank at the time of harvest as was the case of Mendha. In Lakhapur and Lekha overgrazing exhibits its effect on the frequency and density of the saplings.

Regeneration in old growth forest is mediated by canopy gaps [Hartshorn, 1978] and density dependent process [Connell *et al.*, 1984; Harms *et al.*, 2000; Wills *et al.*, 1997]. In the second growth forest, regeneration is more rapid and even due to adequate seed dispersal from nearby mature forest [Chazdon, 2001]. Forest structure, canopy tree composition and processes of seedling, sapling and tree recruitment vary considerably between transitional second growth stands [Guariguata *et al.*, 1997]. Transitional second growth stand of Saigata and old growth stand of Mendha show difference in the recruitment. Species richness of trees above 10 cms in DBH is substantially lower in the Saigata as compared to Lekha and Mendha. Second growth forest has a more homogeneous light environment.

Conclusions

After the analysis of the forest data, the conclusion that can be drawn is that the forest of Saigata, though young in age, has the best potential of growing into a healthy and abundant forest. This forest is not only the best protected one, but also the best maintained one. Mendha, also has managed its forest well. This forest has the added advantage of low population pressure, with per capita forest as high as 4.82 hectares. In case of Lakhapur, though the community has taken steps towards protection, the deteriorating state of the forest is visible in the absence of new plantation. The protection provided could not prove to be efficient enough without cooperation from the formal institution. Damage done to the Lekha forest due to the over harvesting by the community, does not seem to be irreversible as the natural conditions are in favour of this forest. However, there is a serious and immediate need of intervention by the forest department, if the Lekha forest is to be preserved. The case of Maral Mendha is a unique one. It has almost reached a point of no return. There, as mentioned earlier, are no trees and a very few saplings left.The land is far too degraded to be able to support any substantial vegetation. The decision of the forest department to replant a 25-hectare patch can prove useful only if the community can be educated about the merits of conservation.

REFERENCES

Benitez-Malvido, J. (1998). "Impact of forest fragmentation on seedling abundance in a tropical rain forest" *Conserv. Biol*. 12, 380-389.

Chazdon, Robin L (2001). "Forest in transition: Regeneration processes in second growth and mature lowland rain forest of northeaster Costa Rica". In KN Ganeshaiah, R. Uma Shankar and K.S. Bawa (eds.) *Tropical Ecosystem: Structure, Diversity and Human Welfare. Proceedings of the International Conference on Tropical Ecosystems*. New Delhi. 235-238

Connell, J. H., Tracey, J.G. & Webb, L.F. (1984). "Compensatory recruitment, growth, and mortality as factors maintaining rain forest tree diversity". *Ecol. Mongr*. 54, 141-164.

Curran, L.M.I., Caniago, G.D. Paoli, D. Astianti, M., Kusneti, M., Leighton, C.E. Nirarita, H. & Haeruman. (1999). "Impact of El Nino and logging on canopy tree recruitment in Borneo". *Science* 286, 2184-2188.

Daniels, R.J.R., M. Gadgil, & N.V. Joshi, (1995) "Impact of human extraction on tropical humid forests in the Western Ghats in Uttara Kannada, South India". *J. Appl. Ecol*. 32, 296-309.

Foster, R.B. (1980). "Heterogeneity and disturbance tropical vegetation", In *Conservation Biology: An Evolutionary—Ecological Perspective, 75-92* (Massachusetts, USA).

Ganesan, R., T. Ganesh, M. Soubadra Devy and P. Devidar. (2001). "Regeneration Dynamics of a wet evergreen forest, Southern Western Ghats, India". In Ganeshaiah, R. Uma Shankar and K.S. Bawa (eds.) *Tropical Ecosystem: Structure, Diversity and Human Welfare. Proceedings of the International Conference on Tropical Ecosystems*. KN New Delhi 231-234.

Gascon, C. Williamson, B. and da Fonseca, G. A. B. (2000). "Receding forest edges and vanishing reserves", *Science* 288, 1356-1358.

Ghazoul, Jaboury (2001). "Direct and indirect effect of human disturbance on the reproductive ecology of tropical forest trees". In KN Ganeshaiah, R. Uma Shankar and K.S. Bawa (eds.) *Tropical Ecosystem: Structure, Diversity and Human Welfare. Proceedings of the International Conference on Tropical Ecosystems*. New Delhi. 97-100.

Guariguata, M. R. & Pinnard, M. A. (1998). "Ecological knowledge of regeneration from seeds in neotropical forest trees". *For. Eco. Manage*. 112, 87-99.

Guariguata , M.R., R.L. Chazdon, J.S. Denslow, J.M. Dupuy, & L. Anderson, (1997). Structure and floristics of secondary and old-growth forest stands in Costa Rica. *Plant Ecol*. 132, 107-120.

Harms, K., E. S.J. Wright, O. Calderon, A. Hermandez, & E.A. Herre, (2000). "Pervasive density dependent recruitment enhances seedling diversity in a tropical forest". *Nature*, 404, 493-495.

Hartshorn, G. S. (1978). "Treefalls and Tropical forest dynamics". in Tomlinson and Zimmermann (eds.). *Tropical Trees as a Living System*, Cambridge University Press. pp. 616-638

Horne, R. and J. Hickey (1991). "Ecological sensitivity of Australian rainforests to selective logging". *Austr. J. Ecol.* 16,119-129.

Huston, M.A. (1994) *Biological Diversity: The coexistence of Species on Changing Landscape* [Cambridge University Press].

Julieta Benitez- Malvido and Miguel Martinez- Ramos, (2001). "Rainforest fragmentation reduces understorey plant species richness in Amazonia". In KN Ganeshaiah, R. Uma Shankar and K.S. Bawa [eds] *Tropical Ecosystem: Structure, Diversity and Human Welfare.* Proceedings of the International Conference on Tropical Ecosystems, New Delhi, 224-227.

Laurance , W. F., S. G. Laurance, L. V. Ferreira, J. M. Rankin-de-Merona, C. Gascon, & T.E. Lovejoy, (1997). "Biomass collapse in Amazonian forest fragments". *Science,* 278, 1117-1118.

Laurance , W. F., L. Ferreira, J. M. V. Rankin-de-Merona, S .G. Laurance , R.W. Hutchings, & T.E. Lovejoy, (1998a). "Effects of forest fragmentation on recruitment pattern in Amazonian tree communities". *Conser. Biol.* 12, 460-464.

Laurance , W. F., L. Ferreira, J. M. V. Rankin-de-Merona, S.G. Laurance , (1998b). "Rain forest fragmentation and the dynamics of Amazonian tree communities". *Ecology* , 79, 2032-2040.

Laurance , W. F., P. Delamonica, Laurance , Vasconcelos, & T.E. Lovejoy, (2000). "Rainforest fragmentation kills big trees". *Nature* , 404, 836.

Murali, K.S., R. Uma Shankar, KN Ganeshaiah and K.S. Bawa, (1996). "Extraction of non-timber forest products in the forest of Biligiri Rangan Hills, India 2, Impact of NTFP extraction on regeneration , population structure and species composition". *Econ Bot.* 50, 252-269.

Pinard, M.A. & M.D. Swaine, (2001). "Tree regeneration and timber extraction: Patterns, constraints, and research priorities". In KN Ganeshaiah, R. Uma Shankar and K.S. Bawa (eds) *Tropical Ecosystem: Structure, diversity and Human Welfare.* Proceedings of the International Conference on Tropical Ecosystems. New Delhi, 251-256.

Putz , F.E., K. H. Redford, J.G. Robinson, R. Fimbel, & G.M. Blate, (2000). "Biodiversity conservation in the context of tropical forest management", World Bank Environment Department, *Biodiversity series,* Impact studies. Washington, DC.

Scariot, A. (1999). "Forest fragmentation effects on palm diversity in central Amazonia". *J. Ecol.* 87, 66-76.

Shukla, R. P. and S. K. Pandey, (2001). Plant diversity and regeneration in sal forest communities as related to disturbance. In KN Ganeshaiah,

R. Uma Shankar and K.S. Bawa (eds.), *Tropical Ecosystem: Structure, Diversity and Human Welfare*. Proceedings of the International Conference on Tropical Ecosystems. New Delhi, pp 82-86.

Sinha, Aditi and K.S.Bawa, (2001). "Impact of anthropogenic pressures on population dynamics, demography, and sustainable use of forest species in the Western Ghats, India", In KN Ganeshaiah, R. Uma Shankar and K.S. Bawa (eds.), *Tropical Ecosystem: Structure, diversity and Human Welfare. Proceedings of the International Conference on Tropical Ecosystems*, New Delhi, pp. 1001-103.

Sukumar, R., H.S. Suresh, H.S. Dattaraja, & N.V. Joshi, (1998). "Forest biodiversity research in monitoring and modelling: Conceptual background and Old World Case Studies". In Dallmeir, F. & Comiskey, J.A. (eds.) *Man and Biosphere Series*. UNESCO, Paris, New York : The Parthenon Publishing Group.

Wills, C., Condit, R., Foster, R. B. & Hubbell, S.P. (1997). "Strong density and diversity related effects help to maintain tree species diversity in a neo-Tropical forest". *Proc. Nat. Acad. Science* [USA] 94, 1252-1257.

7

LESSONS FROM THE UNCOMMONS

The existence of a political, social, and physical space for learning from and arguing with one another is one important "design principle" that should not be taken for granted.
—Bonnie McCay

The attempt to document the process of institution building in case of four communities was undertaken to investigate the factors leading to the success of collective action, as well as, to look into the reasons leading to inactivity on the part of one community. It can be briefly stated that leadership and coordination between formal and informal institutions are the two factors that are most conducive to sustainable management of common pool resources (CPRs). Living in a democracy, neither is total defiance of formal structure possible, nor is it advantageous to ignore existing (informal) leadership that has been accepted by all under the institutional arrangement prevailing at local level. Sustainable use of forest resources is one of the primary objectives of present National Forest Policy. Decentralization is recognized to be a key to that end. Thus, the 73rd Amendment to the Constitution of India, wherein decentralization is encouraged in the form of transferring some rights and functions to local *gram panchayats*, is a step in that direction. Keeping this in view, the present study primarily deals with successes and failures of community-initiated institutions in sustainable use of forest resources. When coordination between formal and informal institutions is lacking, it invokes two types of reactions: either defiance of the statutory framework or a self-initiated activity, which may not necessarily be contradictory to the broader objective. In either of the situations, both the institutions have to bear a heavy cost. It is commonly seen that usually there are various institutions informally working in each village. Some of these provide incentives to forest users to engage in sustainable use of forest, often resulting in development of the resource. These institutions generally prescribe low discount rate, giving more importance to ecological and bequest value of forests. But there also are institutions, or rather lack of institutions, where members of the community act as individuals (and not as a group necessary for collective action) for their own benefit, irrespective of their impact on surrounding common pool resources (CPRs). Such actions could be compulsive, arising out of poverty and lack of income generating employment opportunities, or could be strategic and consciously taken. Present study incorporates each of these types in its sample of villages.

Fascinating Query

What motivates humans to act in such diverse ways apart from the natural or accidental differences, makes an interesting query. It is fascinating, because many a time communities have to bear substantial cost in the form of time spent, efforts put in, postponement of current consumption etc. for preservation of a resource (in this case the forest resource). Especially in case of communities that are surviving with bare minimum, communities which are officially referred to as 'below poverty line', the cost in terms of 'how much they have to give up' today, for some uncertain benefits in future, is awesome. Saigata village, a sample village in this study, provides one example of this. The whole community of Saigata is so motivated that it is unaffected by the apparently prosperous neighbouring community of Maral Mendha. Two decades ago, before entering the JFM agreement with the Forest Department, when Saigata villagers decided to take things in their hands and took a conscious decision to leave their forest alone for natural regeneration, it had many immediate consequences. The women had to travel long distances for fetching fuel and fodder, or buy it at a price in the nearby market. Men had to stay awake at nights to protect the forest from intruders. Sometimes even women had to join the men in guarding the forest. There were several occasions when they were physically harmed by timber smugglers while protesting against thefts.

Mendha community had to fight a long battle with none other than the "strong" Forest Department, to assert its traditional rights over the forest produce. To this day, Mendha community has to carry on its income-generating forestry activities without the knowledge of Forest Department. For making that possible, it has to take each decision unanimously, which is very costly (Ghate, 2000). Still, the whole community is under constant fear of possible vindictive action that the department is legally authorized to take. Lakhapur community has not been very successful in improving the quality of its forest mainly because it has refused to accept any help from the Forest Department, labelling it as corrupt department with malafide intentions. It is paying heavily for its collective action in the form of foregone developmental funds. All these are examples of the inevitable when there is lack of coordination between formal and informal institutions.

Some Common Assumptions on the Commons

While looking for reasons of failure and successes, one encounters some hypotheses that have been tested in the literature on commons. Two factors that often find mention in this regard are: 'homogeneity of the user group', and 'proximity from markets'. In Indian context, caste system being an important factor that determines social hierarchy is considered to be leading to heterogeneity. Whether homogeneity is conducive for collective action or is it the other way round, is a much argued question. Both the dimensions have been proved and disproved through various studies. (Blair, 1996; Cernea, 1998, 1999; Wade, 1994; Sarin, 1996;

Molinas, 1998; Agrawal, 1994, 1997; Quiggen, 1993; Johnson and Libecap, 1982; Vedeld, 1997) Similarly, proximity to larger towns, markets, and urbanized population is assumed to influence success of collective action in natural resource management. However, no conclusive statement can be made about the hypothesis regarding 'proximity' of larger towns. Earlier studies in this context have shown that proximity of larger towns (of markets, urban population etc. thereby) can work both ways: induce or mar collective action in CPR management. Collective action may fail because of demonstration effect, possibilities of making quick money, or influence of foreign values etc. On the other hand, proximity can also prove to be conducive, if it enhances awareness of the community regarding conservation. It may also mean that the village becomes accessible to non-governmental organizations. But in this study Saigata, where collective action has succeeded and Maral Mendha, where it has not, are two neighbouring villages, equally distant from Bramhapuri market. If the impact of proximity to larger towns can be the reason for deforestation and expansion of agricultural land in Maral Mendha, why was the negative impact deterred in Saigata? Was it because of committed leadership? Leadership that could successfully nurture traditional culture that believes in deferred gratification as against modern/western culture that promotes immediate gratification due to its emphasis on material values.

It is also difficult to conclusively establish the necessity/role of 'homogeneity' in collective action, from this study. While Mendha is a homogeneous community, Saigata is highly heterogeneous, and in both the cases, the community has succeeded in protecting its resource through collective action. Even in Lakhapur where there is limited success, the caste mix of the population is substantial. It substantiates the argument that neither size nor homogeneity is a variable with uniform effect on the likelihood of organizing and sustaining self-governing enterprises (Ostrom, 1999).

However, our study has brought forth 'leadership' as an important factor in the success of collective action in CPR management.

Role of Leadership

"Leadership potential is critical. Both traditional leadership, represented by influential persons from the local area, and modern leadership, represented by college graduates, can play an important role" (Dick, Raju, Gulati, 2002). The process of consensus building, organizing, and maintaining collective action, involves considerable transaction costs that leadership can reduce. Individuals may be influenced by leaders because they may have more faith in information provided by them. If the leadership is from within the community, the members have the opportunity of watching their leader from close quarters. They can immediately know if she/he is disproportionately advantaged by the actions of the whole community. A charismatic leader can give greater assurances to the followers because they believe that the leader has wide influence.

Emergence of a leader as well as his acceptance by a community depends on various factors. This leadership can be informal/traditional or formal i.e. the one provided by larger constitutional structure like *Panchayat* system or the Forest Department, as is the case in the present context. According to leading sociologist, Srinivas (1972), sources of 'power', as manifested in leadership, could be traditional due to caste dominance, economic dominance, or numeric dominance. Numeric dominance often leads to political clout and thereby access or closeness to power. And this affinity to power can be used for economic advantage, and resultantly in more political power. Nature of group activity, for which leadership is being considered, is equally important for determining leadership. In case of intellectual activity, minority leadership is probable. But for any democratic activity, numbers do matter. It is only when the democratic process is vitiated, that the money power gains more importance than numbers or intellects.

Leadership in Saigata refutes this theory to some extent. The leader here belongs to a low caste, a caste that does not outnumber other castes in the village, and he certainly does not belong to the economically dominant class as well. It is true that in Vidarbha region, to which all the five case studies belong, Mahar community (to which the local leader belongs) although considered socially backward, has been historically close to political centre. There had been a practice of appointing *kotwals* (who were the lowest link of the bureaucratic set-up for revenue collection) from this caste. There also were *malgujars / zamindars* (local land holders recognized by the British for helping in collection of land cess) belonging to socially backward castes. These *malgujars* belonged to economically upper strata of the population. Many *kotwals* in Vidarbha region belonged to Mahar caste, and were respected for being 'government officials'. But the leader in Saigata is neither a *malgujar,* nor a rich man. The fact that the problems arising out of scarcity of forest resource were just as severe for the leader as were for the people, helped avert the possible ego clashes within the community. In addition to this, Mr. Khobragade never assumed the role of a leader. He propagated social outlook in an unconventional and informal manner. He also tried to coordinate formal leadership (coming from ownership) of Forest Department with the purpose of giving authenticity to its efforts of regeneration and protection activities, and for ensuring sustainable use of forest. His leadership came into lime light much later due to publicity given to him by some urban NGOs. But his initial success in the form of total acceptability had its roots in sincerity of purpose, indigenous methods of spreading the message, and clear, transparent functioning. He has never hesitated to share all the vital information with his fellow villagers, so as to facilitate their upliftment, thus creating mutual trust. Since he believes in active participation by the leader, he has put in conscious efforts to involve himself with all forest related activities that the village takes up under his leadership. Values adopted by this community are compatible with the broad objective of sustainable use. This complementation of value systems is what is meant by *gram swaraj,* the Gandhian concept of decentralized governance.

In case of Mendha, emergence of leadership from within,

was easier. This tribal community, being homogeneous, traditionally accepts leadership arising from within the community, as the traditional *panchayat* is quite powerful. Yet, it is equally true that there are many villages in the region that are totally tribal, but have not come together for a common cause. Success of leadership in the form of Mr. Devaji Tofa, though cannot be totally subscribed to the fact of 'traditionality'; it certainly played an important role. Practice of taking all the decisions unanimously in the general body meetings, not restricting collective action to forest management alone but expanding it to all the community related spheres, acceptance of and regard for modern norms, and most importantly, maintaining the component of flexibility in the local/informal institution are some of the important factors that have contributed in the success of traditional leadership in case of Mendha. Democratic values seem to be inherent in this community. There is a fine blending of traditional values and modernity, which is an indicator of political modernization. Involvement of women and all other members of the community in decision-making, a significant aspect of democratic decentralization, was accepted in Mendha much before it was accepted as a governance principle by the Constitution of India. However, there is no coordination between this informal institution and the formal institution i.e. the Forest Department. In fact there is defiance to some extent on part of the Mendha community because of the adamant bureaucratic attitude of the department. Actions of the local institution are independent of the directions given by the formal leadership. This remains true even after the village entered into the arrangement with Forest Department through the Joint Forest Management Programme. But, the objective of sustainable forest use being common to both the institutions, the impact of independent actions is not contradictory.

Lakhapur village is the only village in the sample that has been led from 'outside'. The community is highly influenced by a charismatic religious leader, Tukaram Dada, who appealed to the community in the name of judicial rights as well as in the name of social and religious responsibilities. The methods used are also religious: daily prayers, moral classes for children in the community etc. Unlike Saigata, Lakhapur has adopted a defiant approach. But this defiance is qualitatively different from that of Mendha. It neither cooperates with the Forest Department, nor does it accept financial assistance by joining programmes like JFM etc.

Good leadership is often stated to be the requirement for successful collective action, yet it cannot alone achieve high levels of institutional performance (Israel, 1987). Most practitioners believe that leadership is very critical, if not the only critical factor for successful development of institutions. An important question is how to provide 'leadership' when it is not available within the community. It is difficult to find leaders who have vision, have the ability to influence others, who are benevolent and accept equity an as important criterion in the working of groups. One wonders if there are substitutes to leadership to reduce dependence of institutional development on the existence of acceptable leadership. The real question is: are there ways to facilitate the emergence and development of such leadership? We may be able to find some means of doing so if we examined leadership from the point of view of what they do, with which intention they do it, and how they convince the community into following them. An attempt to find out what leadership does to facilitate collective action, may help us find alternative means of ensuring success. In case of Lekha only after the failure of formal *panchayat* leadership in delivering the common good, alternate leadership evolved and is being gradually accepted.

A leader is generally perceived by the followers to have 'vision' for providing solutions to local problems. This alone can make individuals with varying, even diverse interests, permit a leader to influence them. Lack of such leadership in Maral Mendha resulted in creation of a chaotic situation. Formal leadership that could have come from local governing bodies or the Forest Department did not fulfil the local leadership demand.

Anomie and Deviant Behaviour

A clash between goals and institutional means is a common experience in case of economies under transition. It is true at macro/national level as well as at micro/community level. Conflict between cultural goals and institutional means results in unbalanced condition of social structure putting pressure upon individuals or communities. This can provoke behaviour that is contrary to social expectations. This is the state of anomie or 'norm lessness. According to Merton (1968), cultural goals and institutionalized means are two important elements of social and cultural structures influencing the individuals or communities. Ordinarily, the goals are integrated constituting a frame of 'aspirational reference' (discussed later in context of 'reference group behaviour theory'). These goals are pursued through institutionalized means that are approved by the society though they may not be the most efficient means. Disjunction between cultural goals and institutionalized means is how Merton defines anomie. If we consider 'sustainable forest use' as a socially or culturally accepted broad goal, and the democratically accepted and created Forest Department as the institutionalized means, the sample villages reflect different stages or modes of anomie. Different communities in a larger society adapt differently to cultural goals and institutional means. Merton has pointed out positive modes of adaptation to cultural goals i.e. conformity and innovation. While conformity provides stability and continuity to social system, innovation accepts the goals but rejects the means. The negative modes of adaptations are ritualism and retreatism. Ritualism involves abandoning or scaling down or even neglecting the goals, while over-emphasizing institutional means. Retreatism occurs when an individual assimilates both the cultural goals and institutional means but does not achieve success, resulting in retreat or alienation. Rebellion rejects prevailing values, substituting them by new values and different/new means.

We can apply Merton's concept of typology of 'Modes of Individual Adaptations' to the five communities studied here in the following manner.

Table 7.1: A Typology Of Modes Of Adaptation

Sr. No.	Modes of adaptation	Cultural goals	Institution-alized means	Apt village under study
I.	Conformity	+	+	Saigata
II.	Innovation	+	-	Lakhapur
III.	Ritualism	-	+	Mendha (to some extent)
IV.	Retreatism	-	-	Lekha (to some extent)
V.	Rebellion	+/-	+/-	Maral Mendha

Note: (+) signifies 'acceptance'; (-) signifies 'rejection'; and (+/-) signifies 'rejection of prevailing values and substitution of new values'.

This is a broad placement of villages in the categories mentioned above, as no village can fit in any category totally. There always is some overlapping. As has been mentioned earlier, there is no conflict in Saigata between formal and informal institutions. This is a state of total conformity that has resulted into good forest regeneration. The community is getting benefits of various schemes implemented by the Forest Department. On the other hand, Lakhapur has innovated its own institutional means. Although Lakhapur accepts the goal of sustainable forest use, it doesn't accept the role of Forest Department. In case of Mendha, the goals of both formal and informal institutions are same; the role of Forest Department has been very puritanical. This inflexibility gives rise to conflicts in practice. For example, the guidelines of JFM mention that the community can reap the benefits of forest produce like grass for fodder, etc., only after it has guarded the forest for five years. Mendha community had been protecting its forest for a long time and had adopted rotational grazing. Shortly after signing the MOU, a plot of grass was ready to be cut. The department objected to using the grass, saying that the community had to wait for five years, as was the provision under the JFM. Lekha, to some extent, is an example of 'retreatism'. For a long time Lekha community was taking forest supply for granted and did not bother for sustainable forest use. The department overlooked the exploitation of the forest by the community. It is only now, that the community is waking up to the need of conservation and is also requesting the department to initiate JFM programme. In contrast, Maral Mendha is a total rebellion. It neither accepts the goal of sustainable forest use, nor does it bother about given institutional structure. In connivance with the Forest Department officials, the individuals in the community have encroached on the forest land to increase land under cultivation. This is a typical example of anomie at both the levels of leadership.

'Reference Group Behaviour Theory', an important contemporary theory in Sociology, explains a possible source of influence determining why a person or a group behaves in a particular way. Reference group is an ideal that determines the thoughts, conduct, values, status, and the behaviour pattern of others. In case of all the five sample villages, there is no reference group that has influenced their behaviour. The indigenously emerged institutions in Saigata, Lakhapur, and Mendha are neither related to other groups nor do they have knowledge of any 'ideal' community to influence them. These have come up independently without any prompting. However, reference group behaviour can influence the behaviour of a community in future. What if Maral Mendha becomes the reference group for the coming generation in Saigata and Lakhapur because of the immediate and obvious prosperity, coming from increasing land under cultivation at the cost of forest land? Unfortunately Saigata doesn't become reference group for 'Maral Mendha' type of communities because there are no immediate benefits there. From this point of view it is essential that the conservation efforts of the communities like Saigata, Mendha, and Lakhapur also result in material benefits. Direct stakes, yielding results in short term (better than the provisions in JFM), need to be created in such cases. At the same time coercive action is needed to stop Maral Mendha from becoming a 'reference group'. Otherwise there is an immediate danger of imitation. If these forests are managed (including harvesting) on scientific grounds, it could lead to income generation. Another important issue in this context is that of 'ownership'. At present the three villages are protecting their forest without any direct authority on either the forest produce or on the land. The land ownership as well as rights on all the forest products are with the government resulting in suspicion and uncertainty in the minds of communities that have invested their time and efforts in conserving the resource. They feel that the government can undo all their achievements with a stroke of pen i.e. by changing its policy. Therefore, if these conservation efforts are to be made sustainable as well as more acceptable, tenure rights also need to be transferred to the communities.

This again brings to fore another issue of conflict between wider (national or global) interests against the local (community) interests. Both of these can be served only if there is coordination at micro and macro levels, sharing of responsibility, know-how, and final produce. Local leadership without the backing of a right policy or a formal enforcement agency, can achieve little. "Without some specialists who monitor, record information and interpret the rules in a consistent way, the shared community understanding can so erode that the rules become meaningless. If the specialists are not themselves subject to review by others – including all members of the community – their shared understanding of rules can also disintegrate and be replaced by local despotism. Thus, I would argue that neither community nor enforcers are sufficient. Both are needed and both can enhance the other." (Ostrom, 1992: 351)

Relative Positions on Conservation-Harvesting Frontier

An attempt has been made here to locate the five cases studied here, on conservation-harvesting trade-off curve. Proportion of conservation and harvesting of forest products differ from village to village. For sustainable development of forest resource, conservation and harvesting both are to be properly balanced. Higher level of conservation with very low harvesting will obviously not give better results in the long run, though it may be necessary until a minimum level of forest quality is attained. Similarly, harvesting of forest products without investing into conservation may result in eventual depletion of the resource. Using 'conservation-harvesting' frontier, relative positions of villages in terms of conservation and harvesting are presented here on the basis of qualitative as well as quantitative data collected through IFRI method. Points *A* and *B* are extreme limits of conservation and harvesting respectively. As we move in a downward direction from *A* to *B*, area of harvesting gets expanded. Similarly when we move from *B* to *A* in an upward direction, harvesting operation slows down with increased rate of conservation. Horizontal line C represents the 'critical minimum level'. If a resource goes below this level, its revival is extremely difficult.

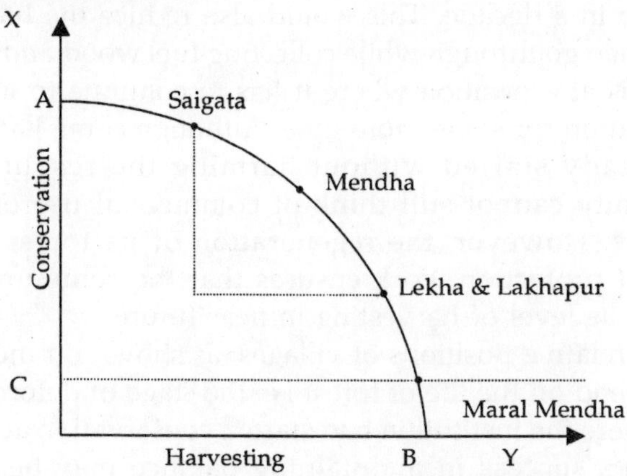

Fig. 7.1: Conservation-Harvesting Frontier

For all the villages, situation relating to conservation and harvesting can be indicated with the help of this diagram. Mendha can be considered as the ideal village in maintaining balance between conservation and harvesting. This would give long-term benefits to the Mendha village in terms of forest resource. For other villages, higher harvesting and lower conservation can prove to be disastrous in the long run. In Lakhapur, unless coordination between local and formal institutions is established, no betterment in the condition of forest can be expected. There are no functional institutions in Lekha and Maral Mendha. In case of Lekha, people are willing to join hands with the Forest Department. Only if the department reciprocates, forest in Lekha can survive. Maral Mendha is in a difficult situation. The forest there has gone below the critical minimum level (C) and has almost reached a point of no return. The only hope of its survival lies in a small fraction of the community members who have not encroached on the land and want the department to intervene, vacate encroachments and replant the area. This group also expressed its wish to build up a nursery in its village to supply saplings for plantation. Women in the community also felt confident that if the department wishes, it can still establish its claim on the forest land, and forest can come up in a decade. This would also reduce the hardships the women go through while collecting fuel wood and fodder. Saigata is at a position where it has to continue to invest in conservation for some more time. Although some harvesting has already started without harming the resource, the community cannot still think of commercial use of forest products. However, the regeneration of its forest due to excellent protection work ensures that the community can increase its level of harvesting in near future.

The relative positions of villages, as shown on the curve, also depend on the life of forest i.e. the stage of deforestation from where the institution had started conservation activities. Mendha's success in maintaining balance may be due to existence of good quality forest right from the beginning of collective action. In case of Saigata, the forest had deteriorated to a great extent and this made harvesting almost impossible in the initial period. It is still on the higher side of conservation.

Factors Conducive to Collective Action

A vast body of literature has come up in last two decades on the subject of commons. After the publication of a number of case studies documenting successes in community-initiated resource management, efforts to identify and isolate factors that are conducive to collective action were taken up. Agrawal

(2002) has integrated multiple variables considered important in 'successful' resource management, drawing from extensive empirical work done by a number of scholars. Wade (1988), from his study of irrigation works in southern India, came up with 14 conditions that facilitate successful management of commons. These include small size, social capital, high dependence of the community on the resource, locally devised access and management rules. Similarly, Baland and Platteau (1996) found small size of a user group, a location closer to the resource, homogeneity among group members, effective enforcement, external aid and strong leadership as significant causal factors for initiating collective action. Nature of the resource, whether it is stationary (Blomquiste *et al*, 1994) or mobile (Moseley, 1999; Naughton-Treves and Sanderson, 1995), storable or not, are some other characteristics that can potentially affect collective action. Work of Ostrom is more elaborate. In the beginning, on the basis of 14 case studies, Ostrom (1990) designed 8 principles i.e. "essential elements or conditions that help to account for the success of these institutions in sustaining the CPRs and gaining the compliance of generation after generation of appropriators to the rules in use" (1990:90).

In brief, the design principles are:

Design Principle 1. Having rules that clearly define who have the rights to use a resource and the boundaries of that resource.

Design Principle 2. Congruence between the rules that assign benefits and the rules that assign costs. And, both types of rules are well matched to local conditions such as soils, slope, number of diversions, crops being grown, and so forth.

Design Principle 3. Collective choice arrangements used to modify the operational rules of regular operation of the resource.

Design Principle 4. Monitoring of rule conformance.

Design Principle 5. Graduated sanctions.

Design Principle 6. Rapid access to low-cost conflict-resolving mechanism.
Design Principle 7. Minimal recognition of rights to organize.
Design Principle 8. Organized activities in multiple layers of nested enterprises.

Later, Ostrom (1999) further suggested attributes of the resource and attributes of the appropriators that are conducive to an increased likelihood that self-governing associations will form. Attributes of the resource are:

1. *Feasible improvement*: Resource conditions are not at a point of deterioration such that it is useless to organize or so underutilized that little advantage results from organizing.
2. *Indicators*: Reliable and valid indicators of the condition of resource system are frequently available at a relatively low cost.
3. *Predictability*: The flow of resource units is relatively predictable.
4. *Spatial extent*: The resource system is sufficiently small, given the transportation and communication technology in use, that appropriators can develop accurate knowledge of external boundaries and internal microenvironments.

And, attributes of the appropriators are given as:

1. *Salience*. Appropriators are dependent on the resource system for a major portion of their livelihood or other important activity.
2. *Common understanding*. Appropriators have a shared image of how the resource system operates and how their actions affect each other and the resource system.
3. *Low discount rate*. Appropriators use a sufficiently low discount rate in relation to future benefits to be achieved from the resource.

4. *Trust and reciprocity.* Appropriators trust one another to keep promises and relate to one another with reciprocity.
5. *Autonomy.* Appropriators are able to determine access and harvesting rules without external authorities countermanding them.
6. *Prior organizational experience and local leadership.* Appropriators have learnt at least minimal skills of organization and leadership through participation in other local associations or learning about ways that neighbouring groups have organized.

Drawing mainly from the work of Ostrom, we have selected six relevant attributes for our study of the five communities to understand their sustainability.

Selected Criterion

Selection of the following criteria was determined on the nature of our enquiry as well as on the basis of our context. Independent variables like size of the user group, tenure rights, dependence on the resource, proximity of the resource to the settlements are held constant. The economic background of all the five communities is also more or less same. The two groups of villages: Saigata, Lakhapur, and Maral Mendha as one group and Mendha and Lekha as another, are equidistant from closest markets. Controlling for their impact thereby, the following attributes were consciously selected on the basis of their relevance in the context of this study. The comparison is not for evaluating the cases and pass value judgements in a strict sense, but to know the likelihood of which community is more likely to sustain its collective action effort. IFRI protocols helped us in collecting abundant qualitative data. Because of the RRA method used in collecting the information, we could capture the overall opinion of each community on difficult aspects like 'social capital' and 'discount rate'. A statistical technique, Multi-criteria analysis was used to analyze the results. The criteria selected for understanding sustainability of the five institutions are as follows:

1. Distribution of Benefits—Assuming that assurance regarding equitable distribution of benefits is necessary for collective action, we wanted to know how equitable is the distribution of benefits arising out of conservation within each community. It is important to note here that in all the villages, forest produce is used/collected by the villagers mainly for self-consumption. The baseline understanding is— each household should get according to its need. An attempt to quantify the forest products used by the households (mainly fuel, fodder and timber for house construction) reveals that the quantity does not vary substantially across the households in the case studies. Therefore, more importance was given to people's perceptions regarding benefit distribution under the institutional arrangement.

2. Conflict Resolving Mechanism—'Infractions are fewer if the sanctions (not necessarily monetary) are imposed by indigenously evolved institutions', is the hypothesis here. Now the question is, was the conflict resolving mechanism in each community developed indigenously or externally? IFRI has a set of questions on incidence of infractions, nature of infractions, rules and punishments related with infractions, and whether these have reduced or increased over a period of time. Through these it was ensured whether the conflict resolving mechanism was indigenous or supplied by an outsider agency.

3. Subscribed Discount Rate—It is believed that cooperation is successful only if the appropriators assign a low discount rate in relation to future benefits. To know this we wanted to know the importance of future benefits over the present for all the communities under study. A low discount rate increases the present value of future income. Calculating the present value of a resource investment using a high discount rate could lead to a shift in the resource use from the future to the present, provided that it does not render the investment economically infeasible. The discount rate is important because it influences the economic efficiency of resource investments and inter-generational equity. This was judged on the basis of attempts made by the community to restrict use of forest products, reduce their needs of forest products, their readiness to pay today both in labour and money to protect the forest for the future, their cultural views of the forest, their awareness of the effects of degradation of forest on other resources etc. The more a community is ready to sacrifice in the present to save the resource for the future, the lower is the discount rate and vice versa.

4. Social Capital—Shared norms, common understanding, and mutual trust can make collective action possible as well as sustainable. Which of the five institutional structures makes best use of the existing social capital? In an attempt to find answer to this question, our enquiry included the activities for which the user group usually came together, homogeneity and heterogeneity of caste, ethnic groups, and religions, unanimity of decision, existence of traditional *panchayat* and/ or level of activity in government promoted *panchayat* and *gram sabha*.

5. Monitoring of resource use—Internal monitoring (by the community) is more effective than monitoring imposed by an external agency, is what we hypothesized. In this regard, it was necessary to know whether the monitoring done by the community is internal or external. Internal monitoring here refers to aspects such as the mode of monitoring, and the timing, as well as the number of guards, and all this to be decided by the community itself, and not imposed by an external agency. This attribute is important because it shows how valuable is the resource for the community and how sincerely they want to guard it from outsiders. For monitoring, the community has to invest a lot more in the form of time and effort.

IFRI has many direct questions dealing with monitoring of resource use. It includes questions like, the number of guards appointed, method of their selection, the types of penalties imposed on infraction(s) for the first time, second time or many times, about the records maintained for penalties, alternative action taken in case of non-payment of fine, restoration of harvesting rights once lost, frequency of general body meetings and the attendance in these meetings etc. Similarly, many questions related to the awareness of the community on the internal or external thefts, about the monitoring by women, the number of guards, and duration of guard duties, the change in the number of thefts etc. were covered, under group discussions.

6. Coordination between formal and informal institutions—The hypothesis here is that the co-ordination between the formal (forest Department) and informal institutions is conducive to collective action. The related research question is— in which community is there coordination between formal and informal institutions? It is often found that the community efforts alone may not prove to be effective if these are not backed by the formal organization. In that case infractions as defined by the informal organization may not be punishable, if the community does not have backing of the government (in this case forest department). Similarly, any formal attempt towards protection of forests unless accepted by the informal organizations, could prove to be equally ineffective. Therefore, it is important to know which of the existing institutional structures is conducive to this kind of co-ordination. The questions covered were mainly to find the coordination between the Forest Department and the local community. For this, questions like the type of activities taken by the Forest Department and the local community together, the type, extent, and quality of coordination between the two, relationship with each other, conflicts if any between the two (lack of it in a village gave more score to it) etc. were asked.

To enable quantification of these qualitative attributes, Multi-criteria analysis was applied by allotting weights to the several questions that were asked using IFRI protocol or during the group discussion. Equal weight (100) was given to each attribute, but the weights to the various questions asked for each attribute varied according to the directness of the questions (Appendix). In this manner, quantitative comparison across the five villages was possible. A conscientious attempt was made to collect data and work out scores for each sample village for each attribute in an objective manner. The scores of the five communities on this basis are as follows:

Table 7.2: Performance of study villages for selected indicators

Sl. No.	*Principles*	*Saigata*	*Mendha*	*Lakhapur*	*Lekha*	*Maral Mendha*
1.	Distribution of benefits	89	89	67	0	0
2.	Conflict resolving mechanism	95	95	56	0	0
3.	Low discount rate	59	48	30	12	3
4.	Social capital	66	65	27	13	5
5.	Monitoring of use	85	73	63	0	0
6.	Coordination between formal and informal	86	44	0	6	0
	TOTAL	480	414	243	31	8

These scores were then used in performing the multi-criteria analysis using the software DEFINITE.

Multi Criteria Analysis

Multi Criteria Analysis (MCA) offers a number of ways of aggregating the data on individual criterion to provide indicators of the overall performance of options. Judgment of the decision maker in establishing objectives and criteria, estimating relative importance weights, and judging the contribution of each attribute to each performance criterion, is a key feature of MCA. This method involves implicit and explicit aggregation of each option's/attribute's performance to form an overall assessment in a form of performance matrix. The three steps in this method are:

1. Preparing a performance matrix.
2. Determining whether criteria are independent of each other or not.
3. Estimate the parameters in terms of the value of its

performance/impact on each of the separate criteria/ attribute.

MCA techniques are used to identify a single most preferred option, to rank options, to short list a limited number of options for subsequent detailed appraisal, or simply to distinguish acceptable from unacceptable possibilities. All MCA approaches make the options and their contribution to the different criteria explicit, and all that is required is the exercise of judgement. For comparing different options' contribution to meet the objective, it requires the selection of criteria to reflect performance in the meeting of the objectives. Each criterion must be measurable, in the sense that it must be possible to assess, at least in the qualitative sense, how well a particular option is expected to perform in relation to the criterion. MCA offers a number of ways of aggregating the data on individual criteria to provide indicators of the overall performance of options. A key feature of MCA is its emphasis on the judgment of the decision making team in establishing objectives and criteria, estimating relative importance weights, and to some extent, in judging the contribution of each option to each performance criterion. The subjectivity that pervades this can be a matter of concern. Its foundation, in principle, is the decision makers own choice of objectives, criteria, weights and assessments of achieving the objectives, although 'objective' data can also be included. MCA, however, can bring a degree of structure, analysis and openness to classes of decisions.

For using qualitative data inputs, it is assumed that reliable and transparent support for decision-making is usually best achieved using numerical weights and scores on a cardinal scale. The two important models in MCA are Linear Additive model and Analytical Hierarchy Process (AHP). In Linear Additive model it is reasonably assumed that the criteria are, preferentially independent of each other and if uncertainty is not formally built into the MCA model, then the simple linear Additive evaluation model is applicable. It shows how an option's values on the many criteria can be combined into

Results of the Multi-criteria Analysis

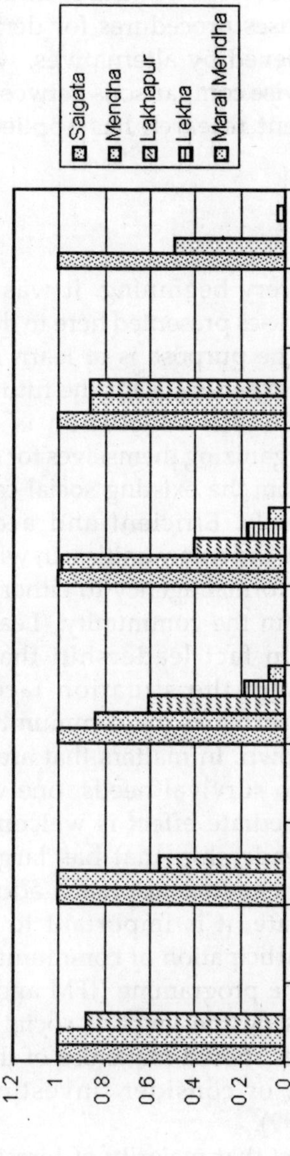

	Distribution of benefit	Conflict resolving	Low discount rate	Social Capital	Monitoring of use	Coordination
Saigata	1	1	1	1	1	1
Mendha	1	1	0.84	0.99	0.86	0.49
Lakhapur	0.88	0.56	0.61	0.4	0.86	0
Lekha	0	0	0.18	0.17	0	0.03
Maral Mendha	0	0	0.07	0.07	0	0

attributes

one overall value. AHP also develops a Linear Additive model but in its standard format, uses procedures for deriving the weights and the scores achieved by alternatives, which are based, respectively, on pair-wise comparisons between criteria and between options. Present research has applied Linear Additive model.

Lessons Learnt

As was mentioned in the very beginning, it was not our purpose to compare the five cases presented here to determine the successes and failures. The purpose is to learn from the communities and from their experiences for the future policy implications. The most obvious conclusion is that the communities are capable of organizing themselves for common purpose. They can benefit from the existing social capital, or can even build it consciously. Efficient and acceptable leadership may not exist in all the communities, in which case it is the responsibility of the formal agency to either provide one, or develop it from within the community. 'Leadership' need not be charismatic. In fact leadership that is not completely well versed with the situation faced by a community, may not be acceptable as the community would not consider him/her as its own. In matters that are of day-to-day concerns and close to survival needs, one who can provide solutions with immediate effect is welcome more than the typical political leadership that has large 'mass appeal'. In the wake of participatory approach adopted by the Forest Department of late, it is important to keep in consideration the fact that participation of communities does not depend so much upon the programme (JFM and similar programmes), as it depends on the level of social capital. Therefore, the formal agencies should make use of the social capital that already exists, or consider 'investing' in it (Grootaert and Narayan, 1999).

One cannot ignore the fact that majority of forest area in India belongs to the state. Although the results of the study reveal self-initiated efforts are more effective as well as sustainable, in the given situation it is highly unlikely that communities will gain exclusive rights over forests and be the sole decision-makers. At the same time one cannot expect that a very large number of communities will take up forest conservation on their own without any tenure right. The study shows that 'Saigata' did not become a role model for Maral Mendha. To assume that 'Maral Mendha' will never become a model for the younger generation of Saigata, would be highly risky. Nor can the society afford to wait for all the forest-dependent communities to take up protection of the remaining meagre forests in India. Therefore, the best solution would be a coordinated effort. Forest department needs to take the lead here, identify and respect the communities that are already protecting or are ready to participate in forest management; be a catalyst, provider of expertise as well as funds wherever required, and supply 'appropriate leadership' where the need arises. We need to accept that self-initiated efforts cannot become a pattern or a rule. Yet, these can prove to be most efficient in resource management if groomed properly by the state. Coordinated efforts of the two institutions have the maximum potential of being feasible as well effective on a wider scale.

REFERENCES

Agrawal A. (2002). "Common Resources and Institutional Sustainability", in Ostrom, Dietz, Dolsak, Stern, Stonich and Weber (eds.), *The Drama of the Commons,* Washington DC: National Academy Press,

Agarwal, B. (1994). "Gender and Command Over Property: A Critical Gap in Economic Analysis and Policy in South Asia", *World Development*, 22(10), 1455-1478.

Agarwal, B. (1997). "Environmental Action, Gender Equity, and Women's Participation. *Development and Change*, 28, 1-44.

Baland, Jean-Philippe and Jean-Marie Platteau (1996). *Halting Degradation of Natural Resources: Is There a Role for Rural Communities?* Oxford: Clarendon Press.

Blair, H.W. (1996). "Democracy, Equity and Common Property Resource Management in the Indian Subcontinent", *Development and Change*, 27, 475-499.

Blomiquist, William, Edella Schlager, Shui Yan Tang, and Elinor Ostrom.

(1994). "Regularities from the field and possible explanations" in Elinor Ostrom, Roy Gardner, and James Walker (eds.), *Rules, Games, and Common-Pool Resources* Ann Arbor: University of Michigan Press, pp. 301-16.

Cernea, M. (1988). " Land Tenure Systems and Social Implications of Forestry Development Programs (Pakistan)", in L.P. Fordman & J.W.Bruce (eds.) *Whose Trees? Proprietary Dimensions of Forestry*. Boulder, CO: Westview Press.

Cernea, M. (1989). "User Groups as Producers in Participatory Afforestation Strategies, World Bank Discussion Papers No. 70, The World Bank, Washington DC.

Ghate, R. (2000) "The Role of Autonomy in Self-Organizing Process: A Case Study of Local Forest Management in India", working paper. Bloomington: Indiana University, Workshop in Political Theory and Policy Analysis.

Grootaert, C., and Narayan, D. (1999). *Local Institutions, Poverty and household welfare in Bolivia,* Washington, DC: World Bank.

Isreal (1987), *Institutional Development: Incentives to Performance,* Baltimore, USA: The John Hopkins University Press.

Johnson R.N. & Libecap (1982), Contracting Problems and Regulations: The Case of Fishery", *American Economic Review*, 72(5), 1005-1023.

Krishna, A. (2001), "Moving from Stock of Social-Capital to the Flow of Benefits: The Role of Agency", *World Development*, 29(6), 925-943

Merton (1968) *Social Theory and social Structure,* New York: The Free Press.

Molinas, J.R.(1998), "The Impact of Inequality, Gender, External Assistance and Social Capital on Local Level Collective Action", *World Development*, 26(3), 413-431.

Moseley (1999) " New Ideas, Old Institutions: Environment, Community, and State in the Pacific Northwest", unpublished Ph.D Dissertation, Department of Political Science, Yale University.

Naughton-Treves, Lisa and Steven Sanderson. 1995. Property, politics and wildlife conservation. *World Development,* 23(8): 1265-75.

Ostrom, Elinor, (1990a). *Governing the Commons: The Evolution of Institutions for collective action*. New York: Cambridge University Press.

Ostrom, Elinor (1990b). "A framework for analysis of self organizing and self-governing CPRs" In Elinor Ostrom *Governing the Commons: The evolution of Institutions for Collective Action*.

Ostrom, Elinor (1992), "Community and the Endogenous Solution of Commons Problems", *Journal of Theoretical Politics*, Vol: 4, No. 3, July 1992, pp 343-351

Ostrom, Elinor. (1999). "Institutional Rational Choice: an Assessment of the Institutional Analysis and Development Framework." In Paul A. Sabatier, ed. (1999) *Theories of the Policy Process*. Bolder, CO: Westview

Press, pp 35-71.

Sarin, Madhu (1996). "From Conflict to Collabouration: Institutional Issues In Community Management" In Poffenberger, M. and B. McGean, (eds.) *Village Voices, Forest Choices*.

Srinivas, M.N. (1972) *Social Change in Modern India*, New Delhi: Orient Longman Limited.

Meinzen-Dick, Raju & Gulati (2002), "What Affects Organizations and Collective Action for Managing Resources? Evidence from Canal Irrigation Systems in India", *World Development*, Vol. 30, No. 4, pp 649-666, 2002

Quiggen, J. (1993), "Common Property: Equality and Development", *World Development*, 21, 1123-1138.

Vedeld, T. (1997), "Village Politics: heterogeneity, Leadership and Collective Action among Fulani of Mali", Unpublished Doctoral Dissertation, Agricultural University of Norway.

Wade, R. (1988) 1994, "*Village Republics: Economic Conditions for Collective Action in South India*", San Francisco, CA: Institute for Contemporary Studies Press.

Appendix

IFRI Forms	*Information Collected*
Site Overview Form	Site overview map, local wage rate, local units of measurement, exchange rates, recent policy changes, interview information
Forest Form	Size, ownership, internal differentiation, product harvested, uses of products, master species list, change in forest area, appraisal of forest condition.
Forest Plot Form	Tree, shrub, and sapling size, density, and species within 1, 3, and 10-metre plots for a random sample of plots in each forest.
Settlement Form	Socio-demographic information, relation to markets and administrative centres, geographic information about the settlement.
User Group Form	Size, socioeconomic status, attributes of specific forest user group.
Forest Association Form	Institutional information about forest association (if one exists at the site), including association's activities, rules structure, membership, record keeping.
Forest-User Group Relationship Form	Products harvested by user group from specific forests and their uses
Forest Product Form	Details on three most important forest products (as defined by the user group), temporal harvesting patterns, alternative sources and substitutes, harvesting tools and techniques, and harvesting rules.
Non-harvesting organization Form	Informations about organizations that make rules regarding a forest(s) but do not use the forest itself, including structure, personnel, resource mobilization, and record keeping.
Organizational Inventory and Inter-organizational Arrangements Form	Information about all organizations (harvesting or not) that relate to a forest, including harvesting and governance activities.

BIBLIOGRAPHY

Adams, B. D. Brockington, J. Dyson, B. Vira (2001). 'Common Choices: Policy potions for Common Poll Resources', Personal Communications, Department of Geography, University of Cambridge.

Agrawal Arun, (2000). "Group Size and Successful Collective Action: Forest Management Institutions in Nine Himalayan Villages." In Clark Gibson, Margaret McKean, and Elinor Ostrom (eds.) *People and Forests: Communities, Institutions, and Governance,* Cambridge, MA: MIT Press.

Agrawal, Arun (2001). "Common Property Institutions and Sustainable Governance of Resources". *World Development* Vol.29: No.10: 1649-1672.

Agrawal, Arun (2002). "Common Resources and Institutional Sustainability", in Ostrom, Dietz, Dolsak, Stern, Stonich and Weber (eds.), *The Drama of the Commons,* Washington DC: National Academy Press.

Agrawal, Arun and Ostrom, Elinor, (1999). "Collective Action, Property Rights, and Devolution in Forest and Protected Areas Management." W99-11, Working Paper, Workshop in Political Theory and Policy Analysis, Indiana University, Bloomington.

Agrawal, A. and C. Gibson (2001). 'The Role of Community in Natural Resource Conservation', in Arun Agrawal and Clark Gibson (eds.) *Communities and the Environment,* New Jersey: Rutgers University Press, pp. 14.

Agarwal, B. (1994). "Gender and Command Over Property: A Critical Gap in Economic Analysis and Policy in South Asia", *World Development,* 22(10), 1455-1478.

Alagh, Y.K. (1999). "Macro Economic and Sectoral Policies and Forestry Sector in India–Role of the World Bank". Draft. World Bank document.

Andersen, K.E. (1995). "Institutional Flaws of Collective Forest

Management". *Ambio*, 24(6): 349-53.

Arora, H., and Khare, A. (1994). "Experience with the Recent Joint Forest Management Approach". Paper prepared for the IN workshop on India's Forest Management and Ecological Revival, New Delhi, February 1994.

Balakrishnan, Mundanthra, and Ndhlovu Dora E. (1992). "Wildlife Utilization and Local People: A Case Study in Upeer Lupande Game Management Area, Zambia." *Environmental Conservation* 19(2): 135-143.

Baland, Jean-Philippe and Jean-Marie Platteau (1996). *Halting Degradation of Natural Resources: Is There a Role for Rural Communities?* Oxford: Clarendon Press.

Bates, R. H. (1988). "Contra Contractarianism: Some Reflections on the New Institutionalism". *Politics and Society* 16:387-401.

Benitez-Malvido, J. (1998). "Impact of forest fragmentation on seedling abundance in a tropical rain forest" *Conserv. Biol.* 12, 380-389.

Bishop, K. P. Andrian, and L. Warren (1995). "Protected for Ever? Factors Shaping the Future of Protected Areas Policy." *Land Use Policy*, 12(4): 291-305.

Blair, H.W. (1996). "Democracy, Equity and Common Property Resource Management in the Indian Subcontinent", *Development and Change*, 27, 475-499.

Blomiquist, William, Edella Schlager, Shui Yan Tang, and Elinor Ostrom. (1994). "Regularities from the field and possible explanations", in Elinor Ostrom, Roy Gardner, and James Walker (eds.) *Rules, Games, and Common-Pool Resources*, Ann Arbor: University of Michigan Press, pp. 301-16.

Borrini-Feyerabend, G. (1996). "*Collabourative Management of Protected Areas: Tailoring the Approaches to the Context, Issues in Social Policy*", Gland (Switzerland):IUCN.

Bromley, Daniel W., David Feeny, Margaret McKean, Pauline Peters, Jere Gilles, Ronald Oakerson, C. Ford Runge, and James Thomson (1992), (eds.) *Making the Commons Work: Theory, Practice, and Policy.* San Francisco, CA: ICS Press.

Burman, Roy B.K. (1985). "Issues in Environmental Management Cantering Forest, Role of Tribal Communities." *South Asian Anthropologist*, 6(1): 41-48.

Cernea, M. (1988). " Land Tenure Systems and Social Implications of Forestry Development Programs (Pakistan)", in L.P. Fordman

& J.W. Bruce (eds.), *Whose Trees? Proprietary Dimensions of Forestry*. Boulder, CO: Westview Press.

Cernea, M. (1989). "User Groups as Producers in Participatory Afforestation Strategies, World Bank Discussion Papers No. 70, The World Bank, Washington DC.

Chaudhary, Ashok (1996) "Conservation, Conflict and Struggle: Rajaji National Park." In K.G. Iyer, (ed.) *Sustainable Development: Ecological and Socio-cultural Dimension*. New Delhi: Vikas Publishing House.

Chazdon, Robin L. (2001). "Forest in transition: Regeneration processes in second growth and mature lowland rain forest of northeaster Costa Rica".In KN Ganeshaiah, R. Uma Shaankar and K.S. Bawa (eds.), *Tropical Ecosystem: Structure, Diversity and Human Welfare*. Proceedings of the International Conference on Tropical Ecosystems. New Delhi, pp 235-238.

Chopra, K. (1995). "Forest and other sector: Critical role of Government policy." *Economic and Political Weekly*, Vol. 30 (25): 1480-1482.

Connell, J. H., Tracey, J.G. & Webb, L.F. (1984). "Compensatory recruitment, growth, and mortality as factors maintaining rain forest tree diversity". *Ecol. Mongr*. 54, 141-164.

Curran, L.M.I., Caniago, G.D. Paoli, D. Astianti, M., Kusneti, M., Leighton, C.E. Nirarita, H. & Haeruman (1999). "Impact of El Nino and logging on canopy tree recruitment in Borneo". *Science*, 286, 2184-2188.

Daniels, R.J.R., M. Gadgil, & N.V. Joshi, (1995). "Impact of human extraction on tropical humid forests in the Western Ghats in Uttara Kannada, South India". *J. Appl. Ecol.*, 32, 296-309.

Datta, S. and Varalakshmi, V. (1999). "Decentralization: An effective method of financial Management at the grassroots (evidence from India)." *Sustainable Development*, Vol. 7 (3): 113-120.

Dore, E. 1996. "Capitalism and Ecological Crisis: Legacy of the 1980s", in Helen Collinson (ed.), *Green Gurrillas*, London: Latin American Bureau (Research and Action) Ltd.

Duraiappah, A.K. (1998). "Poverty and Environmental Degradation: A Review and Analysis of the Nexus." *World Development*, 26(12): 2169-70.

Fisher, R.J. (1989). "Indigenous Systems of Common Property

Forest Management in Nepal", Working Paper No.18, *Environment and Policy Institute*, East-West Centre, Honolulu.

Fortin, M.J. and Gagnon, C. (1999). "An Assessment of Social Impacts of National Parks on Communities in Quebec, Canada." *Environmental Conservation* , Vol. 26: 3, pp. 200-211.

Foster, R.B. (1980). "Heterogeneity and disturbance tropical vegetation", in *Conservation Biology: An Evolutionary—Ecological Perspective, 75-92* (Massachusetts, USA).

Fujita, Hayami & Kikuchi (1999). " The Conditions for Collective Action for Local Commons Management: The Case of Irrigation in the Philippines", Workshop in Political Theory and Policy Analysis, September 13th 1999.

Gadgil, M. and Fikret Berkes. (1991). "Traditional Resource Management Systems." *Resource Management and Optimization*, 8(3-4): 127-41.

Gadgil, M. and R. Guha (1992). *"This Fissured Land: An Ecological History of India"*. New Delhi, India: Oxford University Press.

Gadgil, M. and M.D. Subhash Chandra (1992). "Sacred Groves." *Indian International Centre Quarterly*, 19(1-2): 183-87.

Gadgil, M. (1998). "Conservation: Where are the People?" In Ravi (ed.) The Hindu Survey of Environment 1998 , *The Hindu*, Chennai Edition, India, p. 107-137.

Gaikwad, V. (1995). "Tribal and the Tiger: A Case Study of the Kanha National Park." Centre for Management in Agriculture, Ahmedabad, India.

Ganesan, R., T. Ganesh, M. Soubadra Devy and P. Devidar (2001). "Regeneration dynamics of a wet evergreen forest, southern Western Ghats, India". In K.N. Ganeshaiah, R. Uma Shaankar and K.S. Bawa (eds.), *Tropical Ecosystem: Structure, diversity and Human Welfare.* Proceedings of the International Conference on Tropical Ecosystems, New Delhi, pp. 231-234.

Gaspart, Frederic and Jean-Phillippe Platteau (2001). 'Institutional and the efficient management of Environmental Resources'. Personal Communication.

Gascon, C. Williamson, B. and da Fonseca, G. A. B. (2000). "Receding forest edges and vanishing reserves", *Science*, 288, 1356-1358.

Gazon, L. (1997) "Institutional Structure and the Effectiveness of Integrated Conservation and Development Projects: Case Study from Madagascar" *Human Organization*, Vol. 56, No. 4.

Ghate, Rucha (1992). *Forest Policy and Tribal Development.*, New Delhi, India: Concept Publishing Company.

Ghate R. and Kulkarni, M. (1998). "A Major Setback to Participatory Efforts: A Story of Tragic Success", *Tiger Link*, Vol. 4, No. 2, September 1998.

Ghate, R. (1998a). "Andhari Tiger Sanctuary (Maharashtra): A Case Study for People's Participation in the Management of Protected Area", *The Indian Forester*, Vol. 124, No.10, 853-860.

Ghate, R. (1998b). "A Modern Sacred Grove". *Wastelands News* (February-April): 33-35.

Ghate, Rucha and M. Kulkarni (1998). "Tribal Turmoil." *Down To Earth*, 7(3): 38-39.

Ghate, R. (2000a). "The Role of Autonomy in Self-Organizing Process: A Case Study of Local Forest Management in India", Working Paper. Bloomington: Indiana University, Workshop in Political Theory and Policy Analysis.

Ghate, R. (2000b). "Global Gains at Local Costs: Imposing Protected Areas: A Case Study from India", Working paper. Bloomington: Indiana University, Workshop in Political Theory and Policy Analysis.

Ghazoul, Jaboury (2001). "Direct and indirect effect of human disturbance on the reproductive ecology of tropical forest trees". In K.N. Ganeshaiah, R. Uma Shaankar and K. S. Bawa (eds.) *Tropical Ecosystem: Structure, Diversity and Human Welfare.* Proceedings of the International Conference on Tropical Ecosystems, New Delhi, pp. 97-100.

Gibson Clark, Margaret McKean, and Elinor Ostrom, (2000). *People and Forests: Communities, Institutions, and Governance.* Cambridge, MA: MIT Press.

Gibson, Clark. (Forthcoming). "Dependence, Scarcity, and the Governance of Forest Resources at the Local Level in Guatemala." In Joanna Burger, Richard Norgaard, Elinor Ostrom, David Policansky, and Bernard Goldstein, (eds.) *The Commons Revisited: An Americas Perspective.* Washington, DC: Island Press.

Gichuki, F.N. (1999). "Threats and Opportunities for Mountain Area Development in Kenya." *Ambio*, Vol. 28, Iss. 5, 430-435.

Gillingham, S. and Lee, (1999). "The Impact of Wildlife-related Benefits on the Conservation Attitudes of Local People around the Selous Game Reserve, Tanzania." *Environmental Conservation* , Vol. 26, Iss. 3, pp. 218-228.

Government of India, Ministry of Home Affairs. 1982. *Report on the Committee on Forests and Tribals in India.*

Green Michael J.B. and Paine James. (1997). "State of the World's Protected Areas at the end of the Twentieth Century." Paper presented at the IUCN World Commission on Protected Areas Symposium on 'Protected Areas in the 21st Century: From Island to Networks', Albany, Australia, 24-29 November, 1997.

Guariguata, M. R. & Pinnard, M. A. (1998). "Ecological knowledge of regeneration from seeds in neotropical forest trees". *For. Eco. Manage.* 112, 87-99.

Guariguata , M.R., R.L. Chazdon, J.S. Denslow, J.M. Dupuy, & L.Anderson, (1997). "Structure and floristics of secondary and old-growth forest stands in Costa Rica". *Plant Ecol.* 132, 107-120.

Guha, R. (1983). "Forestry in British and Post-British India: A Historical Analysis." *Economic and Political Weekly,* 18(4): 1882-96.

Guha, R. (1989). "Radical American Environmentalism and Wilderness Preservation: A Third World Critique." *Environmental Ethics,* 11:71-83.

Grootaert, C., and Nararyan, D. (1999). *Local Institutions, Poverty and Household Welfare in Bolivia,* Washington DC: The World Bank.

Haeuber, Richard. (1993). "Development and Deforestation: Indian Forestry and Perspective." *The Journal of Developing Areas* 27: 485-514.

Hames, R. (1987). "Game Conservation or Efficient Hunting", in B.J. McCay and J.M. Acheson (eds.). *The Question of the Commons: The Culture and Ecology of Communal Resources.*

Hardin, G. (1968). 'The Tragedy of Commons'. *Science,* 162(1): 243-48.

Harms, K. E. S.J. Wright, O. Calderon, A. Hermandez, & E.A. Herre, (2000). "Pervasive density dependent recruitment enhances seedling diversity in a tropical forest". *Nature,* 404, 493-495

Hartshorn, G. S. (1978). Treefalls and Tropical Forest Dynamics. In Tomlinson and Zimmermann (ed.) *Tropical Trees as a Living System,* Cambridge University Press, pp. 616-638.

Horne, R. and J. Hickey (1991). "Ecological sensitivity of Australian rainforests to selective logging". *Austr. J. Ecol.* 16, 119-129.

Hurni, H. (1999). "Sustainable Management of Natural Resources in African and Asian Mountains." *Ambio*, Vol. 28, Iss. 5, pp. 382-389.

Huston, M.A. (1994), *Biological Diversity: The Coexistence of Species on Changing Landscape*, Cambridge University Press.

Isreal (1987), *Institutional Development: Incentives to Performance*, Baltimore, USA: The John Hopkins University Press.

Ite, U.E. (1997). "Community Perception of the Cross River National Park, Nigeria", *Environmental Conservation*, 23(4): 351-357.

IUCN, (1994). *1993 United Nations List of National Parks and Protected Areas.* Prepared by WCMC and CNPPA. IUCN, Gland, Switzerland and Cambridge, UK

Jeffery, R., N. Sunder, and P. Khanna, (1998). "Joint Forest Management: A Silent Revolution Amongst Forest Department Staff?", Paper presented at the Workshop on Participatory Natural Resource Management, Mansfield College, University of Oxford, U.K.

Jena, Nalin R. (1996). "National Parks and Sanctuaries vs. People's Rights: Some Issues of Concern." In K.G. Iyer. (ed.) *Sustainable Development: Ecological and Socio-cultural Dimensions* New Delhi: Vikas Publishing House.

Jewitt, S. (1995). "Voluntary and 'Official' Forest Protection Committees in Bihar: Solutions to India's Deforestation?", *Journal of Biogeography*, 22(6):1003-21.

Johnson R.N. & Libecap (1982), "Contracting Problems and Regulations: The Case of Fishery", *American Economic Review*, 72(5), 1005-1023.

Julieta Benitez-Malvido and Miguel Martinez-Ramos, (2001). "Rainforest fragmentation reduces understorey plant species richness in Amazonia". In K.N. Ganeshaiah, R. Uma Shankar and K.S. Bawa (eds.) *Tropical Ecosystem: Structure, Diversity and Human Welfare.* Proceedings of the International Conference on Tropical Ecosystems. New Delhi. pp 224-227.

Kant, Shashi and Roshan Cooke. (1998). "Complementarity of Institutions for the Success of Joint Forest Management: A Comparative Case of Four Villages from India", paper presented at International Workshop on Community-Based

Natural Resource Management, Washington D.C., May, 10-14, 1998.

Krishna Anirudha (2001). "Moving from the Stock of Social Capital to the Flow of Benefits: The Role of Agency," *World Development* 29, No. 6 (June): 925-943.

Krishnaswamy, A. (1995). "Sustainable Development and Community Forest Management in Bihar", *Society and Natural Resources* 8(4): 339-50.

Kulkarni, S.D. (1974). "Over a Century of Tyranny." *Economic and Political Weekly*, 9(10): 389-92.

Kulkarni, S.D. (1979). "Encroachments on Forest Lands—The Experience in Maharashtra." *Economic and Political Weekly*, 19 (45): 1846-49.

Kulkarni, S.D. (1982). "Encroachment of Forests—Government versus People." *Economic and Political Weekly*, 17 (3): 55-59.

Lapiere R.T. and P.R. Fransworth (1949). *Social Psychology*. New York: McGraw Hill Book Co.

Laurance , W. F., S. G. Laurance, L.V. Ferreira, J. M. Rankin-de-Merona, C. Gascon, & T.E. Lovejoy, (1997). "Biomass collapse in Amazonian forest fragments". *Science*, 278, 1117-1118

Laurance, W. F., L. Ferreira, J. M. V Rankin-de-Merona, S .G. Laurance, R.W. Hutchings, & T.E. Lovejoy, (1998a). "Effects of forest fragmentation on recruitment pattern in Amazonian tree communities". *Conser. Biol.*, 12, 460-464.

Laurance, W. F., L. Ferreira, J. M. VRankin-de-Merona, S.G. Laurance , (1998b). "Rain forest fragmentation and the dynamics of Amazonian tree communities". *Ecology*, 79, 2032-2040.

Laurance, W. F., P. Delamonica, Laurance , Vasconcelos, & T.E. Lovejoy, (2000). "Rainforest fragmentation kills big trees". *Nature*, 404, 836.

Leach, M., R. Mearns, and I. Scooners. (1999). "Environmental Entitlements: Dynamics and Institutions in Community-Based Natural Resource Management." *World Development*, 27(2): 225-47.

Lee, P.N. (1993). "Market-State Relations and Environmental Policy: Limit of State Capacity in Senegal." In R.D. Lipschutz and K. Conca. (eds.) *The State and Social Power in Global Environmental Economics*, New York: Columbia University.

Lele, S., and Rao, R.J. (1996). "Whose Co-operatives and Whose Produce? The Case of LAMPS in Karnataka", Paper presented at National Seminar on 'Rediscovering Co-operation', November 19-21, 1996, Institute of Rural Management, Anand.

Lele, S. (1998a). "Godsend, Sleight of Hand, or Just Muddling Through: Joint Forest Management in India", presented at 'Crossing Boundaries', the Seventh Annual Conference of the International Association for the Study of Common Property, Vancouver, British Columbia, Canada.

Lele, S. (1998b). "Why, Who, and How of Jointness in Joint Forest Management: Theoretical Considerations and Empirical Onsights from the Western Ghats of Karnataka", presented at 'Crossing Boundaries', the Seventh Annual Conference of the International Association for the Study of Common Property, Vancouver, British Columbia, Canada.

Libecap, G.D. (1995). 'The Conditions for Successful Collective Action,' in Robert O. Keohane and Elinor Ostrom (eds.) *Local Commons and Global Interdependence: Heterogeneity and Cooperations in Two Domains.* London: Sage Publications pp. 161-190.

Ligon, E., Narain, U. (1999). "Government Management of Village Commons: Comparing Two Forest Policies", *Journal of Environmental Economics and Management*, 37(3): 272-89.

Locke, C. (1999). "Constructing a Gender Policy for Joint Forest Management in India", *Development and Change,* 30(2): 265-85.

McGinnis, Michael, ed. (1999). *Polycentric Governance and Development: Readings from the Workshop in Political Theory and Policy Analysis.* Ann Arbor: University of Michigan Press.

McGinnis, Michael and Elinor Ostrom. (1996). "Design Principles for Local and Global Commons." In Oran R. Young, (ed.) *The International Political Economy and International Institutions,* Vol. II., Edward Elgar, Cheltenham, UK.

McKean, M.A. (1992). "Management of Traditional Common Lands (*Iriaichi)* in Japan". In D. Bromley (ed.) *Making the Commons Work: Theory, Practice and Policy*. San Francisco: Institute for Contemporary Studies, pp. 63-93.

McKean, Margaret A. (2000). "Common Property: What Is It, What Is It Good For, and What Makes It Work?" In Clark Gibson, Margaret McKean, and Elinor Ostrom, (eds.) *People and Forests: Communities, Institutions, and Governance.* Cambridge, MA: MIT Press,

McNeely, K.R., J.A. Miller, R.W. Reid, R.A. Mittermeier, and T.B. Werner, (1988). *"Economics and Biological Diversity: Developing and Using Economic Incentives to Conserve Biological Resources"* Gland, Switzerland: IUCN.

Meinzen-Dick, Raju & Gulati (2002). "What Affects Organizations and Collective Action for Managing Resources? Evidence from Canal Irrigation Systems in India", *World Development*, Vol. 30, No. 4, pp. 649-666, 2002.

Merton (1968). *Social Theory and Social Structure*, New York: The Free Press.

Mitra, Kinshuk (1999). Draft background paper for OED study as mentioned in Preliminary Report: Alleviating Poverty Through Participatory Forestry Development: An Evaluation of India's Forest Development and World Bank Assistance, Jan. 7, 2000.

Mishra, R. (1968). " *Ecology Work Book*" New Delhi: Oxford and IBH Publishing Co., pp. 244.

Molinas, J.R., (1998). "The Impact of Inequality, Gender, External Assistance and Social Capital on Local Level Collective Action", *World Development*, 26(3), 413-431.

Moseley (1999). "New Ideas, Old Institutions: Environment, Community, and State in the Pacific Northwest", unpublished Ph.D Dissertation, Department of Political Science, Yale University.

Mukharjee, Neela (1997). "Why Joint Forest Management (JFM) Failed to Deliver? A Case Study of Arjuni Mouza, Midnapore (West Bengal, India)." *Indian Forester*, 123(6): 546-55.

Murali, K.S., R. Uma Shaankar, KN Ganeshaiah and K.S. Bawa, (1996). "Extraction of non-timber forest products in the forest of Biligiri Rangan Hills, India 2, Impact of NTFP extraction on regeneration, population structure and species composition". *Econ. Bot.*, 50, 252-269.

Nabli, M.K. and J.B. Nugent (1989). "Collective action, Institutions and Development", in Nabli, M.K. and J.B. Nugent (eds.) *The New Institutional Economics and Development: Theory and Applications to Tunisia*. Amsterdam: Elsevier Science Publishers, pp. 95.

Naughton-Treves, Lisa and Steven Sanderson (1995). "Property, politics and wildlife conservation", *World Development*, 23(8): 1265-75.

North, D.C. (1990). *Institutions, Institutional Changes and Economic Performance*, Cambridge: Cambridge University Press.

Olson, M. (1965). *The Logic of Collective Action: Public Goods and the Theory of Groups*. Cambridge, MA: Harvard University Press.

Ostrom, Elinor (1990a). *Governing the Commons: The Evolution of Institutions for Collective Action* . New York: Cambridge University Press.

Ostrom, Elinor (1990b). "A framework for analysis of self organizing and self-governing CPRs" In Elinor Ostrom, *Governing the Commons: The Evolution of Institutions for Collective Action*. New York: Cambridge University Press.

Ostrom, Elinor (1992a). *Crafting Institutions for Self-Governing Irrigation Systems*, San Francisco, CA: ICS Press.

Ostrom, Elinor (1992b). "Policy Analysis of Collective Action and Self-Governance" In William N. Dunn and Rita Mae Kelly (eds.) *Advances in Policy Studies since 1950, Policy Studies Review Annual*, Vol. 10, pp. 81-119.

Ostrom, Elinor (1992c). "Community and the Endogenous Solution of Commons Problems", *Journal of Theoretical Politics*, Vol. 4, No. 3, July 1992, pp. 343-351.

Ostrom, Gardener & Walker (1992). "Covenants with and without a Sword: Self-Governance is Possible", *American Political Science Review*, Vol. 86, No. 2, June 1992.

Ostrom, Elinor, Roy Gardner, and James Walker (1994). *Rules, Games, and Common Pool Resources.* Ann Arbor : University of Michigan Press.

Ostrom, Elinor (1994). "Self-Organizing Resource Regimes: A brief report on a decade of policy analysis", *Policy Currents*, Vol. 4, No. 3, August 1994.

Ostrom, Elinor (1997). "Self Governance and Forest Resources" Conference on 'Local Institutions for forest management: How can Research make a Difference', CIFOR, Bogor, Indonesia, November 19-21. Http://www.cgiar.org/cifor.

Ostrom, Elinor (1998). "A Behavioral Approach to the Rational Choice Theory of Collective Action", *American Political Science Review*, Vol. 92, No. 1, March 1998.

Ostrom, Elinor (1999a). "Coping with Tragedies of the Commons." *Annual Review Political Science*. 2:493-535.

Ostrom, Elinor. (1999b). "Institutional Rational Choice: An Assessment of the Institutional Analysis and Development.

Framework. In Paul A. Sabatier, ed. (1999) *Theories of the Policy Process*. Bolder, CO: Westview Press, 35-71.

Ostrom, E. (1999c). "Formulating the Commons" In Joanna Burger, Richard Norgard, Elinor Ostrom, David Plicansky, and Bernard

Goldstein, (eds.) *The Commons Revisited*, Washington D.C: Island Press.

Ostrom, Elinor. (1999d). "Self Governance and Forest Resources". Occasional Paper No. 20, Centre for International Forestry Research, Bogor, Indonesia.

Ostrom, Burger, Field, Norgaard & Policansky (1999). "Revisiting the Commons: Local Lessons, Global Challenges", *Science*, Reprint Series, 9th April 1999, Vol. 284, pp. 278-282.

Parker, A.N. (1995). "Decentralization: The way forward for rural development?," World Bank Policy Research Working Paper 1475.

Pathak, Neema (1998). "Lessons from Natural Resource Conservation from a Community Initiative: The Case of Mendha Village, Maharashtra", Seminar on Eco-documentation, Wildlife Institute of India, Dehra Dun, 26-27 November, 1998.

Pathak, N. and Kothari, A. (1998). "Sharing benefits of wildlife conservation with local communities: Legal implications." *Economic and Political Weekly*, Vol. 33 (40): 2603-2610.

Pattanaik. B.K., and Dutta, S. (1997). "JFM in South-West Bengal – A Study in Participatory Development", *Economic and Political Weekly*, 32(50): 3225-32

Pinard , M.A. & M.D. Swaine (2001). "Tree regeneration and timber extraction: Patterns, constraints, and research priorities". In K.N. Ganeshaiah, R. Uma Shaankar and K.S. Bawa (eds.) *Tropical Ecosystem: Structure, Diversity and Human Welfare.* Proceedings of the International Conference on Tropical Ecosystems, New Delhi, pp 251-256.

Poffenberger, M. (ed) (1990). *Keepers of the Forest: Land Management Alternatives In Southeast Asia*, West Hartfort, CT: Kumarian.

Poffenberger, M. and B. McGean, (eds) (1996). "*Village Voices, Forest Choices: Joint Forest Management in India*". Delhi: Oxford University Press.

Poteete, A. and Ostrom (2002). "An Institutional Approach to the Study of Forest Resources", Forthcoming in *Human Impacts on Tropical Forest Biodiversity and Genetic Resources*, ed. John Poulsen Centre for International Forestry Research (CIFOR). New York: CABI Publishing.

Putnam, Robert D. (1993) *Making Democracy Work: Civic Traditions in Modern Italy*, USA: Princeton University Press.

Putz, F.E., K.H. Redford, J.G. Robinson, R. Fimbel, & G.M. Blate,

(2000). "Biodiversity conservation in the context of tropical forest management", World Bank Environment Department, *Biodiversity Series*, Impact Studies. Washington, DC.

Quiggen, J. (1993). "Common Property: Equality and Development", *World Development*, 21, 1123-1138.

Ramnathan, U. (1996). "Displacement and the Law", *Economic and Political Weekly*, Vol. 31 (24): 1486-1491.

Ravindranath, M.H., P. Sudha, and K.M. Indu (1998). "Participatory Forestry Indian Experience in Community Forestry and JFM." Working Paper. Centre for Ecological Sciences, Indian Institute of Science, Bangalore.

Ribot, J.C. (1993). "Market-State Relations and Environmental Policy: Limit of State Capacity in Senegal." In R.D. Lipschutz and K. Conca (ed.) *The State and Social Power in Global Environmental Economics*. New York: Columbia University.

Richards, M. (1997). "Common Property Resources Institutions and Forest Management in Latin America." *Development and Change*, 28 (1): 95-117.

Richards, M. (2000). "Community Forest Management: The Nepalese Experience." *Down To Earth*, 8(19): 30-46.

Rondinelli, D. (1981). "Government Decentralization in Comparative Perspective: Theory and Practice in Developing Countries." *International Review of Administrative Science*, 47.

Sarin, Madhu (1996). "From Conflict to Collaboration: Institutional Issues in Community Management" in Poffenberger, M. and B. McGean, (eds.) *Village Voices, Forest Choices*. Delhi: Oxford University Press, pp. 165-203.

Sarin, M. (1995). "Joint Forest Management in India: Achievements and Unaddressed Challenges." *Unasylva ,180*, Vol. 46 (1): 30-36.

Scariot, A.(1999). "Forest fragmentation effects on palm diversity in central Amazonia". *J. Ecol.*, 87, 66-76.

Sekhar, N.U., A.M. Ulrik, and N. Shanmugratnam (1996). "Park Management, Land Use Patterns and People's Perceptions: The Case of Desert National Park, Rajasthan, India." Presented at "Voices form the Commons," the Sixth Annual Conference of the International Association for the Study of Common Property, Berkeley, CA, June 5-8, 1996.

Sharpe, B. (1998). "First the Forest: Conservation, Community and

Participation in Southwest Cameroon." *Africa*, Vol. 68 (1): 25-45.

Shukla, R. P. and S. K. Pandey, (2001). "Plant diversity and regeneration in sal forest communities as related to disturbance" in K.N. Ganeshaiah, R. Uma Shaankar and K.S. Bawa (eds.) *Tropical Ecosystem: Structure, Diversity and Human Welfare.* Proceedings of the International Conference on Tropical Ecosystems, New Delhi, 82-86.

Singh, C. (1986). *Common Property and Common Poverty. India's Forests, Forest Dwellers and the Law.* Delhi: Oxford University Press.

Singh, S. and Khare, A. (1993). *"People's Participation in Forest Management"*, Wastelands News (May-July).

Sinha, Aditi and K.S.Bawa, (2001). *"Impact of anthropogenic pressures on population dynamics, demography, and sustainable use of forest species in the Western Ghats, India"*, in K.N. Ganeshaiah, R. Uma Shaankar and K.S. Bawa (eds.). Tropical Ecosystem: Structure, Diversity and Human Welfare. *Proceedings of the International Conference on tropical Ecosystems.* New Delhi, pp. 1001-103.

Smoke, P. (1993). "Local Government Fiscal Reform in Developing Countries—Lessons from Kenya." *World Development*, 21 (6) 901-923.

Srinivas, M.N. (1972). *Social Change in Modern India*, New Delhi: Orient Longman Limited.

State of India's Environment – First Citizens' Report, 1982 (1996). Centre for Science and Environment, New Delhi.

Steins & Edwards (1999). "Collective Action in Common-Pool Resource Management: The Contribution of a Social Constructive Perspective to Existing Theory". *Society and Natural Resources*, 12: 539-557, 1999.

Sudha P., P.V. Rekha, V.S. Gunga, S. Patagar, M.B. Naik, K.M. Indu, and M.H. Ravindranath.(1998). "Community Forest Management and JFM: An Ecological, Economic, and Institutional Assessment in Western Ghats, India." Working Paper. Centre for Ecological Sciences, Indian Institute of Science, Bangalore.

Sukumar, R., H.S. Suresh, H.S. Dattaraja, and N.V. Joshi, (1998). Forest biodiversity research in monitoring and modelling: Conceptual background and Old World Case Studies. In

Dallmeir, F. and Comiskey, J.A. (eds.), *Man and Biosphere Series.* UNESCO, Paris, New York: The Parthenon Publishing Group.

Sundar, N. (2000). "Unpacking the 'Joint' in Joint Forest Management", *Development and Change,* 31(1): 255-79.

Tang, Shui Yan. (1992). *Institutions and Collective Action: Self-Governance in Irrigation Systems.* San Francisco: ICS Press.

Thompson, J. (1995). "Participatory Approaches in Government Bureaucracies: Facilitating the Process of Institutional Change", *World Development* 23 (9): 1521-34.

Townsend, Amy K. (1992). "Protected Areas as Common Property and India's Sundarbans National Park." Presented at "Inequality and the Commons," the Third Annual Conference of the International Association for the Study of Common Property, Washington D.C., Sept. 17-20, 1992.

Vedeld, T. (1997), "Village Politics: Heterogeneity, Leadership and Collective Action among Fulani of Mali", Unpublished Doctoral Dissertation, Agricultural University of Norway.

Venter, A K. and Breen Charles M. (1998). "Partnership Forum Framework: Participative Framework for Protected Area Outreach." *Environmental Management,* Vol. 22, No. 6: 803-815.

Verrier Elwin, (1994), *The Aboriginals.* London : Oxford University Press.

Voelekar, J.A. (1893). *Report on the Improvement of Indian Agriculture.* London: Eyre and Spottiswoode.

Wade, R. (1988). 1994, *Village Republics: Economic Conditions for Collective Action in South India,* San Francisco, CA: Institute for Contemporary Studies Press.

Wells, M. (1992). "Biodiversity Conservation, Affluence and Poverty: Mismatched Costs and Benefits and Efforts to Remedy Them." *Ambio,* 21(3): 237-243.

Wells, M. (1995). "Biodiversity conservation and local people's development aspirations: New proprieties for the 1990s." Rural Development Forestry Network, Network Paper 18a, Winter 1994-Spring 1995.

Wills, C., R. Condit, R. B. Foster, & S.P. Hubbell, (1997). "Strong Density and Diversity Related Effects Help to Maintain Tree Species Diversity in a Neotropical Forest". *Proc. Nat. Acad. Science* [USA] 94, 1252-1257.

Wily, L. (1999). "Moving forward in African Community Forestry: Trading Power, Not Use Right", *Society and Natural Resources* 12(1): 49-61.

Wunsch, James S. (1999). "Institutional Analysis and Decentralization: Developing an Analytical Framework for Effective Third World Administrative Reforms." In Michael D. McGinnis (ed.) *Polycentric Governance and Development, Readings from the Workshop in Political Theory and Policy Analysis*. Ann Arbor: University of Michigan Press.

Laws and Bills referred to:

Amendment Proposals for the Wild Life (Protection) Act, 1972, Government of India, April, 1997.

Forest Conservation Act, 1980.

Indian Forest Act, 1927.

National Forest Policy, 1988.

Wild Life (Protection) Act (as amended in 1991).

Index